OAKWOOD LIBRARY OF RAILWAY HISTORY, OL168

FROM GLOUCESTER TO LEDBURY
THE DAFFODIL LINE

John Mair

The Ledbury Junction in 1957 after simplification which reduced the double junction to a single line. Access to and from the branch was now confined to the down platform road at Ledbury station. Here a train from Gloucester is about to join the down main line.
Tony Harden collection

THE OAKWOOD PRESS

© John Mair, 2022

ISBN 978-0-85361-764-8

Printed by
P2D Books, 1 Newlands Rd, Westoning, Bedford, MK45 5LD

For
Julia

At the down platform at Ledbury, the staff and railcar W19W prepare for their next journey to Gloucester. *R. E. Toop*

Front cover: On its way to Gloucester diesel railcar No. W19W pauses at Dymock.
R. E. Toop

Published by
The Oakwood Press, 54-58 Mill Square, Catrine, KA5 6RD
Telephone: 01290 551122,   Website: www.stenlake.co.uk

Contents

Preface 5

One   Historical Outline 7
The Canal Age, 7; Coal... and Disappointment, 8; The Newent Diversion and Completion of the Canal, 11; Railway Planning, 12; Converting the Canal to a Railway, 14; Construction between Ledbury and Dymock, 19; Construction of the Newent Railway, 19; Utilising the Canal, 20; Inspection and Opening, 20; Early Losses of Traffic, 22; Gradients, 26; Stations, 27; Halts, 29

Two   Operating The Line 31
Passenger Train Services, 31; Goods traffic, 42; Motive power, 47; Weight and Speed Restrictions, 51; Ledbury Locomotive Depot, 52; Signalling and Single Line Working, 52; *Over Junction, 54; Barber's Bridge, 56; Newent, 57; Dymock, 59; Ledbury Branch Junction, 60; Ledbury,* 61

Three   A Journey From Gloucester in Later Years 63
Barber's Bridge, 65; On to Malswick, 66; Newent, 66; The Gable, and Four Oaks Halt, 70; Dymock, 70; Greenway Halt and the approach to Ledbury: the Town Halt, 72; Ledbury, 75

Epilogue 77
Bibliography 79
Index 80

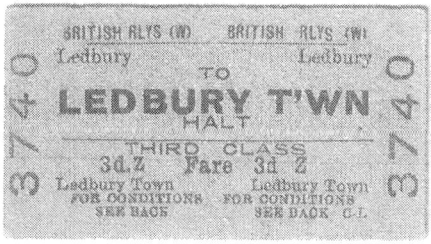

The Railway from Gloucester to Ledbury

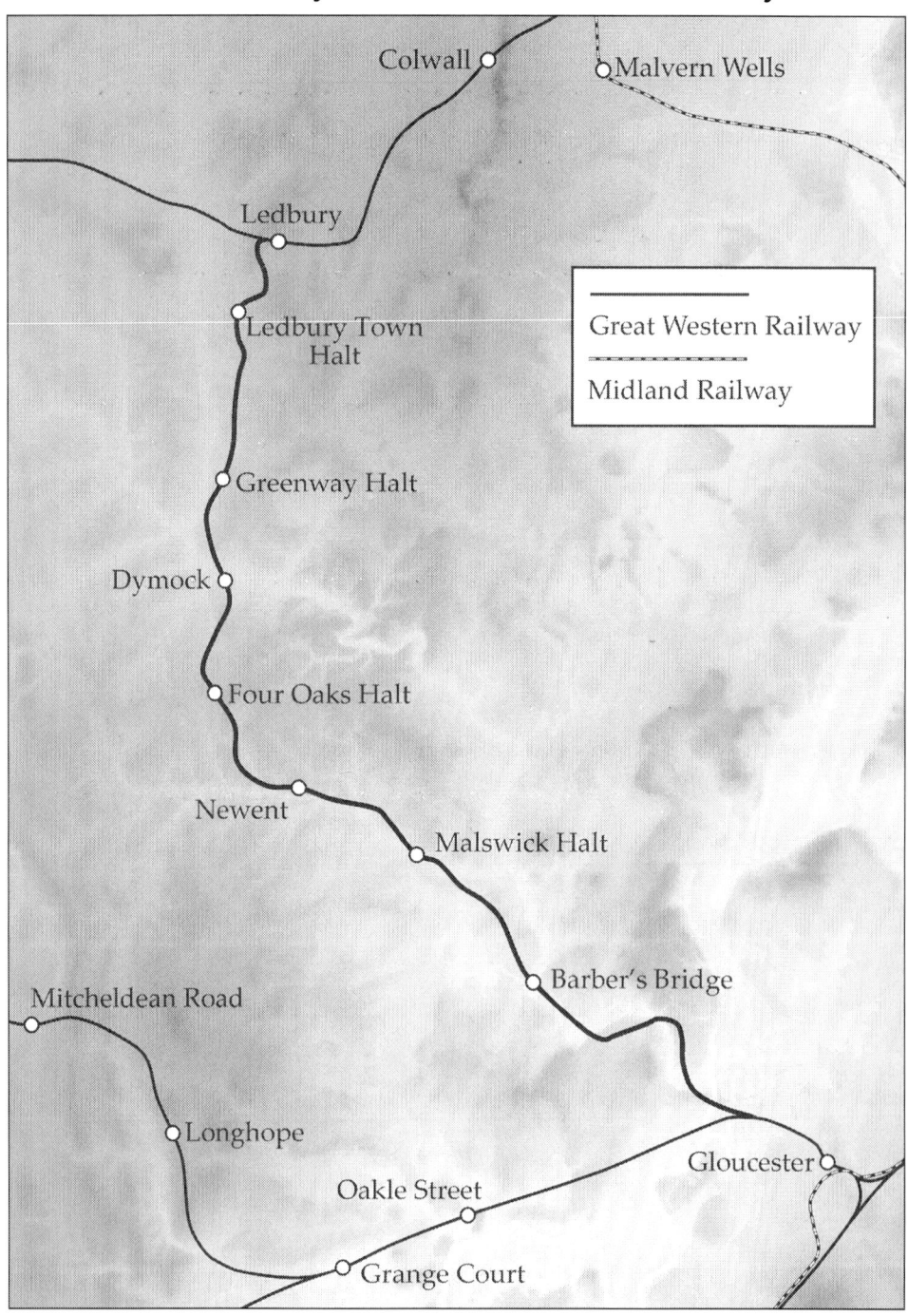

Relief shading contains OS data © Crown copyright and database right 2022.

Preface

Until the high summer of 1959 railway travellers between the cathedral cities of Gloucester and Hereford had a choice of two routes, both of which passed through areas of great natural beauty. The more direct, and quicker, way lay via Ross-on-Wye and traversed parts of the Forest of Dean and of the valley of the River Wye. The second route followed the course of a canal and ran through deepest Gloucestershire to cross the Herefordshire border just to the south of Ledbury, where travellers would change for a train to take them to Hereford itself, or for another to take them to the Malverns, to Worcester and to the West Midlands. Opened in July, 1885, the railway between Gloucester and Ledbury was closed to passengers in mid-July, 1959, and the remaining part of the line, between Gloucester and Dymock, was closed to goods at the end of May 1964.

In preparing this history and description of the railway which ran between Gloucester and Ledbury I have been assisted by several writers. David Bick wrote an account (published in 1979) of the Herefordshire and Gloucestershire canal, and his book included a chapter contributed by John Norris about the Gloucester and Ledbury line. This chapter was an adaptation of an article which Mr Norris had written for *The Railway Magazine* in 1958. In 1994 the Oakwood Press published an enlarged version of Mr Bick's book, and in 2003 the Press published a further edition. Each of these revisions included a contribution by Mr Norris.

In 1985 David Postle of the Kidderminster Railway Museum produced a memoir of the last passenger journey from Ledbury to Gloucester, and in 2002 Leslie Oppitz briefly described the line in his *Lost Railways of Herefordshire and Worcestershire*. More recently, in 2011, Colin Maggs wrote about the railway in his *Branch Lines of Gloucestershire*. I have also made use of the standard reference books, such as the *Regional Histories* series published by David and Charles. Reg Instone, John Lacy and Garth Tilt of the Signalling Record Society have provided valuable information; and John Alsop, Nicky Harden, the Lens of Sutton Association, Colin Maggs, Michael Clemens (www.michaelclemensrailways.co.uk), John Chalcraft of Railphotoprints, and the Great Western Trust have supplied photographs of stations along the line. To all of these writers and sources I am most grateful. Thanks also to Philip Halling, whose photographs on the Geograph website beautifully illustrate the local countryside and in particular the daffodils. Again I am grateful to Lewis Hutton, designer and editor at the Oakwood Press, for his always sage advice, support and encouragement. Lewis also drew the maps on pages 4, 6 and 15.

Stanmore, Middlesex
October 2022

Southern part of the Hereford and Gloucester Canal

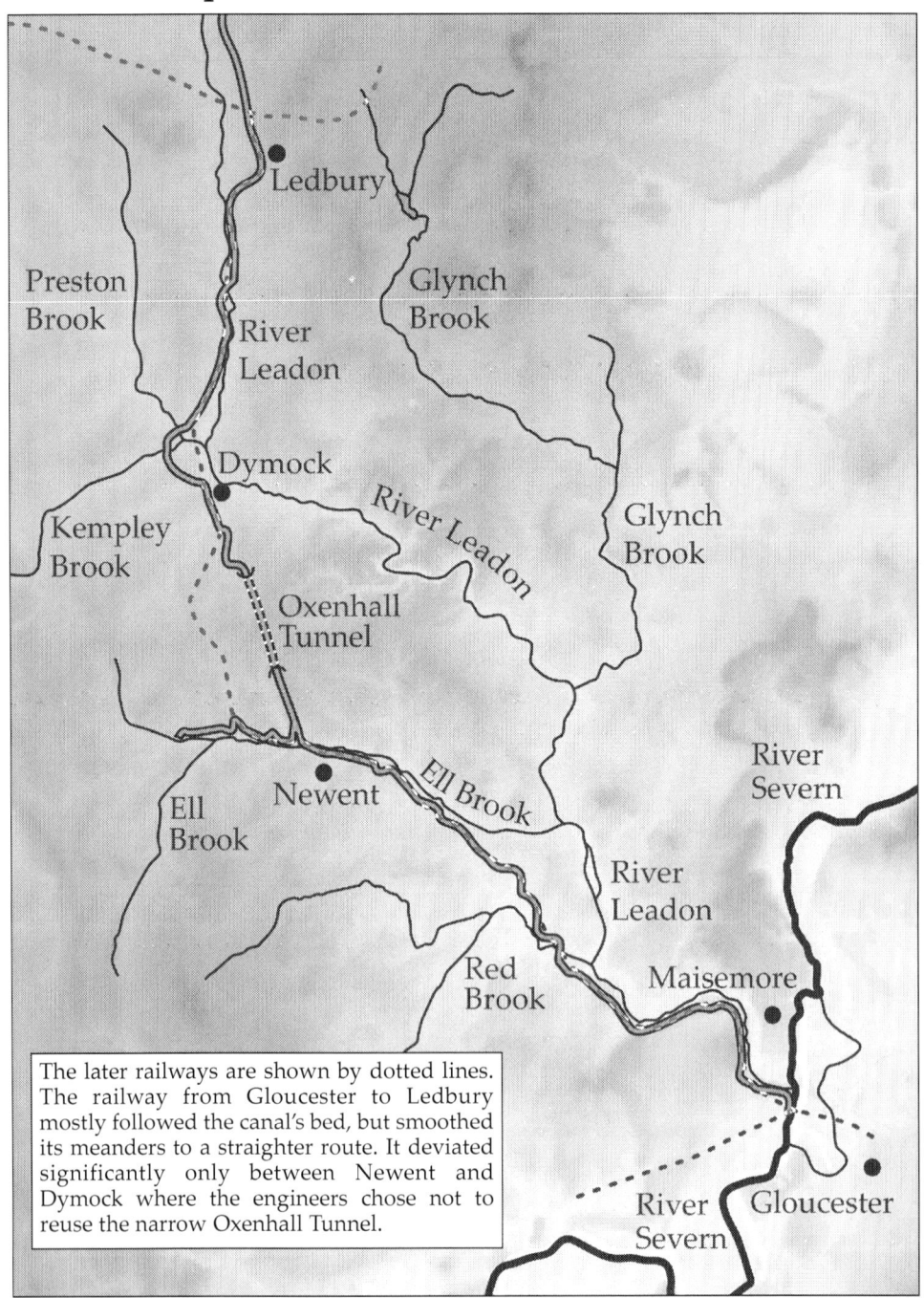

The later railways are shown by dotted lines. The railway from Gloucester to Ledbury mostly followed the canal's bed, but smoothed its meanders to a straighter route. It deviated significantly only between Newent and Dymock where the engineers chose not to reuse the narrow Oxenhall Tunnel.

Relief shading contains OS data © Crown copyright and database right 2022.

Chapter One

Historical Outline

The Canal Age

The rapid development of the canal system in Britain during the last quarter of the 18th century encouraged previously independent and isolated towns and cities to aspire to greater connectivity as a means of acquiring economic and social prosperity. It was in this context that the Herefordshire and Gloucestershire canal was planned.

Proposed by Robert Whitworth in 1777, an early plan in the region envisaged a canal which would begin at Stourport-on-Severn and run south of the Wyre Forest and westwards along the valley of the River Teme. When it reached a point near the village of Orleton it would turn south to reach Leominster and Hereford. Then it would take up an easterly course before moving south to Ledbury, and finally run south-eastwards to return to a navigable part of the River Severn below Tewkesbury. This grand scheme – an arc with both ends touching the River Severn – would connect many inland towns and villages, and provide them with a route to Gloucester and beyond. Alas, the full plan never eventuated, and only an isolated section from Mamble, east of Tenbury Wells, to Leominster was constructed.

Nevertheless, some years later the idea bore fruit in the form of a more modest project to connect the cities of Hereford and Gloucester. In 1789 Richard Hall planned a canal route similar to that contemplated in 1777 for the section from Hereford to Ledbury, and he proposed a continuation from there via the village of Eldersfield to a junction with the River Severn at a point (almost opposite the small settlement of Wainlode Hill) from where craft would be able to continue by the Severn for a distance of about five miles to reach the port of Gloucester. Soon afterwards, however, Hall changed the canal route from Ledbury to a more direct and more easily graded course which would run along the valley of the River Leadon, and which then, crossing the Severn by aqueducts, would lead to Gloucester itself.

Above Gloucester, and near the village of Maisemore, the River Severn divided – and divides – at the 'Upper Parting' into two sections, namely the West Channel and the East Channel, between which lies the Island of Alney, an irregularly shaped and mainly level area, at the southern end of which, the 'Lower Parting', the two Channels reunite.

The main road between Gloucester and South Wales crossed the Island. Alney did, however, form a barrier to any continuation of the

Hereford and Gloucester Canal, and in the event the canal went only as far as the West Channel, which it joined a short distance to the north of the settlement of Over, and about a mile and a half west of Gloucester itself.

The River Leadon – the name means 'Broad Stream' – rises to the south of the town of Bromyard in north Herefordshire. It flows in a generally southerly direction to reach the west side of the town of Ledbury (whose name is derived from that of the river) and to the village of Dymock; and then it turns south-eastwards to pursue a winding course to a confluence with the River Severn near Over. Hall's revised plan was that after its journey eastwards from Hereford to Ledbury the 'main line' of the new canal would generally follow the western banks of the Leadon. From this route a branch would run about three miles westwards to Newent.

Local inhabitants and interests were enthusiastic in their support for this plan, and in 1790 they appointed Josiah Clowes as adviser and engineer. In the following year, 1791, Parliament passed a Bill authorising the establishment of the Herefordshire and Gloucestershire Canal Navigation Company, and empowering it to build the canal and for this purpose to raise capital of up to £75,000 and to borrow up to £30,000. By the summer of 1792 the capital was fully subscribed. Problems, however, lay in store.

Coal... and Disappointment

An inducement to support the Hereford and Gloucester canal (as it became generally known) was the prospect of carrying locally mined coal to Hereford and to other centres. In the vicinity of Newent were a number of coal measures which were outliers of the Forest of Dean coalfield. For many years there had been intermittent attempts to work these deposits, but now more systematic and energetic efforts were to be made. Further, it was now thought, in order to facilitate the efficient carriage of the coal, it would be advantageous to alter part of the route of the canal so that its main line should pass through Newent, thus obviating the need for the three mile branch line to which reference has already been made. Hugh Henshall (a brother-in-law of James Brindley, the renowned canal engineer) drew up a plan for such a revised route. But Clowes strongly advised against this diversion, which would mean that the canal would leave the easily graded Leadon valley and have to negotiate some very high ground to the

HISTORICAL OUTLINE 9

Small prairie locomotive No. 4564 prepares to leave the down platform at Ledbury with the 8.22 pm up train to Gloucester on 6 July, 1955. This is an example of a non auto-fitted engine hauling an auto-trailer (here No. W216W) treated as a conventional coach.
Hugh Ballantyne/Rail Photoprints

Collett 0-4-2 tank locomotive No. 1401 has brought a train from Gloucester into the up platform at Ledbury. *Lens of Sutton Association*

Railcar No. W19W has arrived in the down platform at Ledbury, after the simplification, in 1957, of the junction at the west end of the station. *Lens of Sutton Association*

At Ledbury Town Halt an ex-GWR railcar arrives from Gloucester. Like that at Greenway, this halt was built on the course of the former up line. When opened, the halt was lit by two oil lamps, but in this view an electric light, attached to the upper part of the tall post to the right of the waiting shed, is just visible. Ledbury Town was the most used of the halts but appears here to be dilapidated. *M. Hale/Great Western Trust*

north of Newent. He recommended adherence to Mr Hall's plan and wrote:

> Upon a careful examination of the country from Ledbury to Newent, I find it so much intersected with hills, that it is impracticable to carry the main Line by way of Newent. But a lateral branch to Newent may be as laid down in the plan by Mr. HALL, which, in my opinion, will answer every requisite purpose.

His views were, however, overruled: the proprietors pressed ahead with their plan for a deviation, and they obtained the necessary Parliamentary authorisation in 1793. Meanwhile work on the comparatively simple stretch from Over had begun. Sadly Clowes died in 1795, at the early age of 58, and his post as engineer was backfilled by Robert Whitworth, who may have been, perforce, more complaisant than had been Josiah Clowes towards his wilful proprietors.

In retrospect the outcome was predictable. Previous attempts to mine coal in the area of Newent had met with little success, and the renewed efforts now being made fared little better. The output of coal was limited, and the quality indifferent – so much so that the fuel came to be considered as suitable for lime burning and for little else. In his history of the Hereford and Gloucester Canal, David Bick speaks dryly of the Hill House Colliery at Newent as being 'the least unsuccessful'.

The Newent Diversion and Completion of the Canal

The canal company was, however, by now committed to the Newent route, which entailed the construction of a long tunnel at Oxenhall in the district lying to the north of Newent itself. This tunnel, 2,192 yards in length, gave rise to many problems, mainly that of flooding caused by the springs which were encountered as work proceeded. After much delay and additional expenditure, the tunnel was eventually finished, although it was by no means straight, perhaps because as many as 20 (and possibly more) shafts had been sunk along its length.

On 29th March, 1798, the canal was opened as far as a basin situated close to the Ledbury to Ross road (the present road A449). This supposedly temporary terminus was still some way short of Ledbury town, which in the event was not reached until almost 43 years later, in the year 1841. On 22nd February in that year the first load of canal borne coal arrived at Bye Street, close to the town centre.

The construction of the rest of the canal main line between Ledbury and Hereford lies outside the scope of the present study: suffice it to say here that after numerous vicissitudes it eventually opened in 1845, some 47 years after the completion of the Over to Ledbury Basin section, and in the year in which the Newport, Abergavenny and Hereford Railway obtained its Act to authorise the construction of a railway which would efficiently bring good quality Monmouthshire coal to Hereford. It was much to the credit of a gifted young chief engineer, Stephen Ballard, that the section of the canal between Ledbury and Hereford was built at all: his skill, determination and commitment ensured that the works were carried to completion. The proprietors were aware, however, that in persisting with this new length of canal they were creating an anachronism: they realised that the future of transport lay with railways, which were indeed rapidly becoming a present reality.

Accordingly, the canal owners formed the hope that in time a railway company might purchase parts of their canal for conversion to a line between Ledbury and Hereford. Events were, however, to take a different course. The Worcester and Hereford Railway (W&HR) touched the northern edge of Ledbury (to the west of which the Hereford and Gloucester canal ran), but the promoters of the railway then preferred to pursue a direct route to the northern side of Hereford, rather than adopt the proffered canal route. They were perhaps deterred from using the already completed canal formation because it took a roundabout route to follow the river valleys, and possibly also because of the need to convert the tunnels at Ashperton and at Aylestone (the latter in Hereford). So in the event the W&HR took a direct line westwards from Ledbury, past the villages of Ashperton, Tarrington, Stoke Edith and Withington, to join the Shrewsbury and Hereford Railway at Shelwick Junction, one mile and 56 chains north of Barrs Court station in Hereford.

Railway Planning

In the 1840s proposals for railways in Gloucestershire and Herefordshire had been manifold: one which was to be of at least indirect effect upon the area between Gloucester and Ledbury was for a railway from Gloucester to Hereford. This itself underwent many (and time wasting) changes of plan, but in the event a broad gauge (7 ft 0¼ in.) line was built, running from Grange Court Junction – seven and a half miles south-west of Gloucester and on the main line to South Wales – through the Forest of Dean to Ross-on-Wye, and then on through Fawley and Ballingham in the Wye Valley to Rotherwas and to Hereford itself, and it was eventually opened in June, 1855. The line was converted to the standard gauge in August, 1869.

On 1st July, 1860, the West Midland Railway (WMR) had been formed by an amalgamation of the Oxford, Worcester and Wolverhampton Railway, the Newport, Abergavenny and Hereford Railway, and the (not at that time completed) Worcester and Hereford Railway, which would form a bridge between the two larger railways. The W&HR opened in stages, and services running the entire length of the line from Worcester to Hereford began on 13th September, 1861. Mr W. P. Price of Tibberton Court (which lay to the west of the future Barber's Bridge station) was both the Chairman of the Hereford and Gloucester Canal Company, and the Deputy Chairman of the West Midland Railway.

William Price perceived that with the dismissal of the Hereford and Gloucester canal route between Ledbury and Hereford, and with the direct rail connections now established in 1861 between Ledbury and both Worcester and Hereford, the future of the canal and of communication between Gloucester and Ledbury must inevitably lie with the railways. Upon Price's initiative, a concordat was reached on 17th January, 1862, between the Canal Company, the WMR and the Great Western Railway (GWR) whereby the canal would be managed by a joint committee of the three companies, pending acquisition (which would require Parliamentary approval) of the canal by the railway companies, whereafter the canal company would receive in perpetuity an annual sum of £5,000 in recognition of the transfer of ownership. It was also agreed to seek a provision in the legislation for the canal below Ledbury to be converted to a railway.

The existence of the WMR was to prove short-lived, since on 1st August, 1863, the company was amalgamated with – in reality virtually absorbed by – the Great Western, which accordingly for practical purposes took control of the canal and of its future. The GWR now had no reason to hasten to obtain the legislation envisaged by the 1862 concordat: it was in a position to resist any potential incursions by the expansionist London and North Western and Midland Railways (the latter company enjoyed running powers over the W&HR); and deferment of the legislation would also delay the liability for payment of the annual sum of £5,000 to the Herefordshire and Gloucestershire Canal Navigation Company. Not until 1870 did the GWR obtain powers to finalise its possession of the canal: the Great Western Railway (Hereford and Gloucester Canal Vesting) Act received Royal Assent on 4th July in that year, and the GWR could now acquire the canal outright, together with the liability to pay in perpetuity the annual sum to the canal company. This liability was eventually extinguished when most of the railways of Britain were nationalised under the Transport Act, 1947, and the canal shareholders were rewarded with British Transport stock. In this way, the canal was more profitable to its owners for long after it had ceased to be used as a waterway.

At this point, mention should be made of another scheme in the area. The Worcester, Dean Forest and Monmouth Railway was authorised by an Act of that name, dated 21st July, 1863. It was to run southwards from Great Malvern (on the W&HR) to Newent, and then south-westwards through the Forest of Dean to Mitcheldean, Coleford and Monmouth. A subsequent proposal was for a branch line from Newent to Gloucester, making some use of the canal. A ceremonial start was made by the cutting of the first sod at Speech House, in the heart of the Forest; and some – quite isolated – preparatory earthworks were carried out in open country about two miles to the north-east of Newent. Otherwise, however, little if anything was done, and the project was gradually abandoned. Monmouth received a rail connexion when a line following the River Wye was opened from Ross to May Hill station (close to the centre of Monmouth) on 4th August, 1873.

Converting the Canal to a Railway

Meanwhile, the Herefordshire and Gloucestershire Canal Navigation Company continued to exist, to operate the canal, and indeed – thanks to prudent management and to some well judged rate cutting – to derive some income from it. So far as the generally expected conversion of the canal to a railway was concerned, the Great Western again bided its time, and was content to be a third party when proposals were eventually formulated. Three schemes, including plans for conversion, or for part-conversion, of the canal were put forward in 1872.

One scheme was the Ross, Ledbury and Gloucester Railway. This railway was to start on the Gloucester side of the River Severn, and to cross the East Channel, Alney Island and then the West Channel, before taking up the alignment of the Hereford and Gloucester canal, which it would then follow all the way to Ledbury, from where a new section of railway would run south-westwards to Ross. A Bill to authorise this proposed railway came before Parliament in 1873 but was refused: the Severn Commissioners successfully resisted the Bill on the grounds that the planned bridges over the river would give insufficient clearance for navigation. The disappointed promoters decided not to resubmit their Bill, probably because by then two

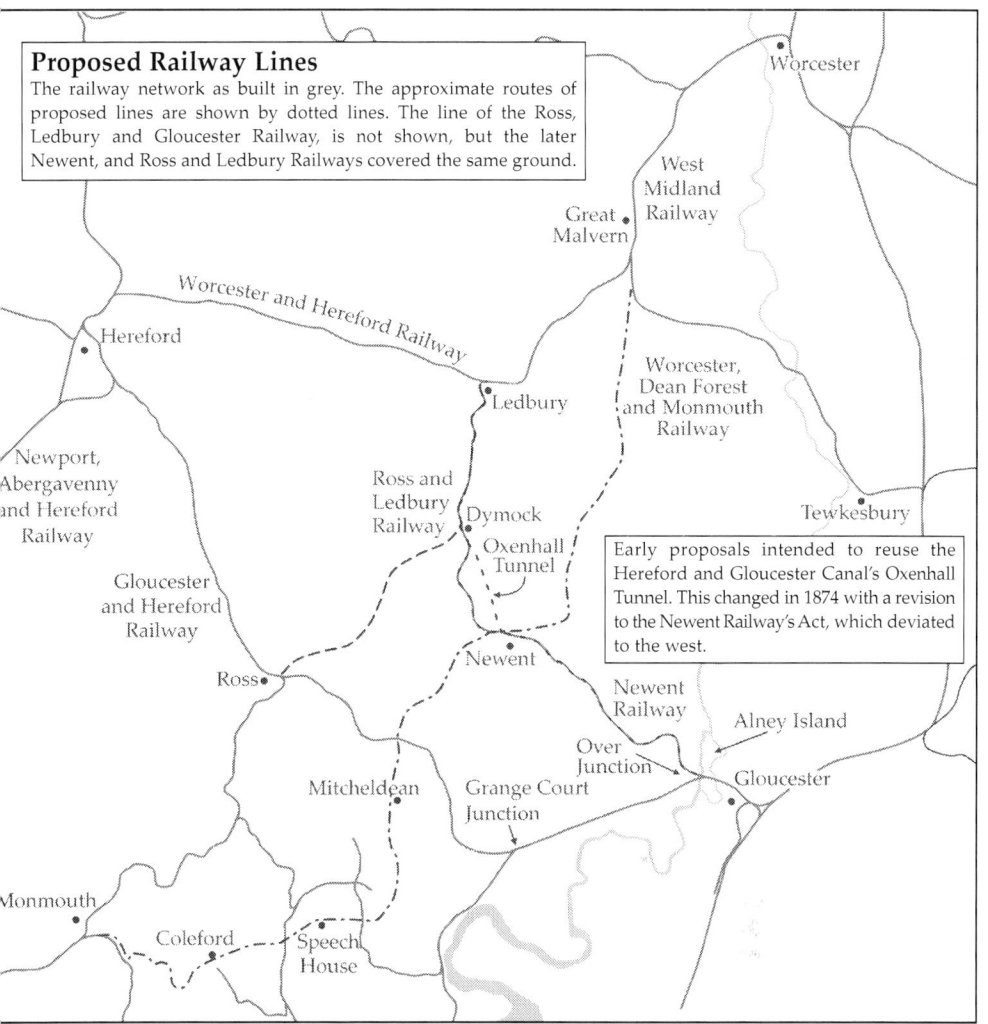

alternative schemes would, when taken together, largely secure the aims of the Ross, Ledbury and Gloucester Railway.

These other two schemes were formally separate from each other, but were complementary and indeed were advanced by the same people. The Ross and Ledbury Railway would begin at a junction with the Worcester and Hereford line at Ledbury, and then follow the course of the canal as far south as Dymock, from where a new line would run south-westwards to Ross, and there join the existing Gloucester to Hereford line via Grange Court.

The complementary scheme was for a nominally separate line, called the Newent Railway, which would run from Over Junction, a short distance to the west of the West Channel of the River Severn and on the

South Wales main railway line, and then, for the most part, follow the alignment of the canal to Dymock, where a junction would be formed with the Ross and Ledbury Railway. When combined, these two railways would achieve most of what was proposed for the dismissed Ross, Ledbury and Gloucester scheme, and, moreover, without the complication of new crossings over the River Severn.

Ross itself would become a railway crossroads. It was already connected by rail to Gloucester (via Mitcheldean Road and Grange Court), and to Hereford, and would be joined by lines, from 1873, to Monmouth and, from 1876, to Chepstow; and now, under the new proposals it would also be reached by trains from Worcester, Malvern, and Ledbury.

Separate Bills for these two closely related railways came before Parliament in 1873, and their passage proceeded without controversy. The Ross and Ledbury Railway and the Newent Railway Bills received Royal Assent on 28th July, 1873, and on 5th August, 1873, respectively.

The Newent Railway Act provided for the railway to traverse the Oxenhall Tunnel, which, as we have seen, ran beneath high ground between Newent and Dymock. Very soon, however, it was realised that the technical difficulties of converting this both narrow and crooked tunnel to railway use would be formidable if not actually insuperable. Accordingly, the company concluded that it would be preferable to go overland and to climb steeply from just outside Newent to a summit at Four Oaks, and then to descend at a similar inclination towards Dymock, and to rejoin the canal formation north of the Oxenhall Tunnel. An Act to authorise this deviation was passed by Parliament on 21st May, 1874. It too was called the 'Newent Railway Act'.

In 1878 Parliament sanctioned extensions of time for the completion both of the Newent and of the Ross and Ledbury Railways. The relevant statutes were the 'Newent Railway' and the 'Ross and Ledbury Railway' Acts, both of which received Royal Assent on 16th April in that year.

Although the enabling legislation was now in place, the schemes themselves lost impetus. Some, but not enough, of the necessary capital was subscribed locally. Accordingly, the Great Western Railway now judged it time to intervene, and offered to take over and to finance the project. The companies themselves remained in being, but their directors retired, to be replaced by GWR nominees under the chairmanship of Sir Daniel Gooch (Chairman of the GWR). Even so, there was little haste, and

HISTORICAL OUTLINE 17

Ledbury station looking east with the junction of the line to Gloucester behind the camera. The station's signal box can be seen on the right. *John Alsop*

Ledbury station photographed from the hill above the tunnel. The station's goods shed is conspicuous, behind it can be seen the signal box. The only visible feature of the railway from Gloucester to Ledbury is the line of the embankment to the left of the main station building. *John Alsop*

Station work at Dymock, as railcar W19 calls on 8th July, 1959, shortly before closure of the line. The William Clarke building on the up platform appears to be still in very good condition. *Author*

At Newent, railcar W19W which regularly worked on the line, waits to depart for Gloucester. The rail access to the goods yard can also be clearly seen. *R. E. Toop*

further delays ensued. The Great Western decided not to proceed with the section between Dymock and Ross-on-Wye: only the section between Ledbury and Dymock would be built, although as a double line (presumably in case the plan to continue to Ross were ever revived).

Construction between Ledbury and Dymock

At last, in November, 1880, the contract to build the double line between Ledbury and Dymock was let, to Messrs Appleby and Lawton. At the beginning of June, 1881, notice was given of the intention 'on and after the 30th day of June 1881 to stop up and close' the portion of the canal between Ledbury and its junction with the River Severn, and the notice also required the removal of all 'barges, boats and other craft accordingly'. After its closure to commercial traffic, the canal was used to transport materials for the construction of the Ledbury to Dymock section, work upon which started in the summer of 1881. The canal itself was of course gradually shortened as the conversion to a railway proceeded southwards. Notable bridges were built across two trunk roads, namely the Hereford Road (now the A438) near Ledbury station, and the Ross Road (now the A449) south of Ledbury. By the end of 1882 work on the Ledbury to Dymock section was well advanced, and attention turned to the construction of the rest of the line.

Construction of the Newent Railway

An invitation to tender for the construction of the line between Over Junction and a junction with the Ross and Ledbury at Dymock was published in the *Gloucestershire Chronicle* on 28th October, 1882, advising that plans and specifications could be obtained from the Engineer, Mr William Clarke, of 45 Parliament Street, Westminster. This William Clarke is now chiefly known for his distinctive design of railway station buildings (see below). The tender notice was signed by G. Cottman, Secretary of the Newent Railway, whose address was given as Padddington Station, the headquarters of the GWR. The contract to construct the line was also awarded to Messrs Appleby and Lawton, who had already been engaged to build the line between Ledbury and Dymock. The line was single, with passing loops at Barber's Bridge, Newent and Dymock, but provision was also to be made in the earthworks and at bridges for the addition of an extra track if required at a later date to make a double line.

Construction began in the following year, 1883, and on 21st July in that year both the *Gloucester Citizen* and the *Gloucester Journal* reported on the very substantial works in progress in the vicinity of Over: they entailed the temporary diversion of the main road (now the A40 trunk road) and the construction of a road bridge under which, after leaving the South Wales main line at Over Junction, the new railway could pass, before running by a cutting through Vineyard Hill to reach the line of the canal, along the course of which much of the railway was to be built.

On 18th December, 1884, the *Gloucester Citizen* could report that the works for the line all the way to Dymock were substantially complete, and that tracklaying had begun. The three station buildings, at Barber's Bridge, at Newent and at Dymock, were built, to the designs of William Clarke, by sub-contractors, Messrs G. Cowdry and Sons of Newent.

Utilising the Canal

The two sections (Ledbury to Dymock, and Over Junction to Dymock) were as far as possible superimposed upon the canal formation, with the notable exception of the Oxenhall deviation, described above. The railway could not of course follow implicitly the complete line of the canal, which included some small but sudden changes of direction and abrupt changes of level at the locks. The replacement of a canal with a standard gauge railway worked by steam locomotives inevitably required a number of adjustments, particularly in matters of curvature and of levels. As it was, the infilling of the canal pounds (the sections between locks) necessitated the deposit of very large volumes of earth. At Newent, the station and the track were partly supported upon arches, which after completion were concealed. Stone taken from the locks was used in the construction of bridges along the lines. The iron girders for these bridges were supplied by Messrs Brattells of Worcester.

Inspection and Opening

On 20th and 21st July, 1885, Colonel Rich inspected, for the Board of Trade, the complete length of line between Over Junction and Ledbury, and reported favourably. The way thus lay open for the commencement of passenger services over the line, and these began on Monday, 27th July, 1885, which was the day of the annual Show of the Gloucestershire Agricultural Society. For this occasion Gloucester station was *en fête*, with Chinese lanterns decorating parts of the building.

The first passenger carrying train left Gloucester for Ledbury on Monday, 27th July, 1885, at 7.10 am, and made a round trip, arriving back in Gloucester at 8.40 am, to the accompaniment of exploding detonators. There was to be a service each day of five trains each way, on weekdays only, and this initial daily pattern of five trains each way was to be long lasting. Until the end of 1916 most of these trains from Gloucester continued to Malvern or Worcester, and vice versa. Freight trains between Gloucester and the West Midlands could now travel by the new line, thus obviating the long and costly detour via Ross and Hereford.

On 1st July, 1892, the Great Western Railway, which had for all practical purposes taken control of the construction and operation of the line, formally acquired the two local companies. The enabling statute was the Great Western Railway Act, which had received Royal Assent on 28th June, 1892.

An early economy in the running of the line was the simplification of Barber's Bridge station. From the outset, this station was a block post: that is, for signalling purposes it divided into two sections the line between Over Junction and Newent. Although, however, provided with a passing loop and with a signal box and with two platforms, one on either side of the loop, Barber's Bridge was never a staff station or crossing place, where trains travelling in opposing directions could pass one another. It

Barber's Bridge station, early in 1909, and looking south towards Gloucester. The down side loop had been removed several years previously, but the platform thus marooned seems to be in still good condition. The road bridge in the middle distance carried the main road from west of Gloucester to Trumpet (Ashperton)(now the B4215).   *John Alsop*

would seem that it might have been a simple matter to upgrade the station to a staff station. But in 1898 the passing loop was taken out of use, and the signal box was closed. Possibly the GWR – ever vigilant for economies – had concluded that for traffic purposes not only was it unnecessary to create new staff sections but also that a block post at this point was superfluous.

This meant that henceforward the block section extended from Over Junction to Newent, a distance of 8 miles and 34 chains. The train staff or tablet for this section was used to allow access to the goods yard at Barber's Bridge. The now redundant down platform there became heavily overgrown, but remained in situ until, and after, the closure of the line.

Early Losses of Traffic

The effect of the completion on 1st August, 1906, of the 'cut-off' railway between Cheltenham and Honeybourne, and the subsequent doubling of the hitherto single line from Honeybourne to Stratford-upon-Avon, was to abstract through traffic from the Gloucester to Ledbury line. When combined with the North Warwickshire line running from Bearley (north of Stratford-upon-Avon) to Tyseley (which line was

The exterior of Ledbury station, seen from the town side. The signal post, half hidden by vegetation, on the left of the bridge marks where the junction for the line to Gloucester is.
John Alsop

opened for goods traffic on 9th December, 1907 and for passengers on 1st July, 1908), the new Honeybourne line provided a more direct and a quicker route both for freight and later for passengers from Gloucester to the Birmingham area and to the West Midlands generally. Passenger trains continued to run over the Newent and Ledbury line to Malvern and Worcester, but, as we have seen, these were withdrawn during the First World War and were never reinstated. Most through freight trains between Gloucester and the West Midlands were diverted to the new route via Honeybourne, and the goods traffic which continued between Gloucester and Ledbury became mainly local in character.

In 1916 the double line between Dymock and Ledbury was reduced to single track, the ostensible reason (as so frequently in cases of this kind) being that the rails were required for use in France. The reality was, often, that the owning company was able to use lightly worn rails for renewals upon its own system. The double junction with the Worcester to Hereford main line at Ledbury did, however, survive, and a small signal box, 'Ledbury Branch Signal Box', was installed in January, 1917, at a point where the by now single line from Dymock became double before forming the junction with the main line west of Ledbury station. This box lasted until 1925, when the double junction was repositioned so as to be slightly nearer the station, and the points could be motor operated. The entire layout at Ledbury was then brought under the control of the signal box situated at the east end of the down platform.

Much later, in 1957, the double junction at the west end of the Ledbury station layout was replaced by a single lead junction, with the result that trains proceeding from and to Gloucester could from then on use only the down main line and the down platform.

The story of the Gloucester to Ledbury line after the first decade of the 20th century was one of gradual contraction. It was, however, only gradual, because the line was better placed than many other rural lines to resist competition from the roads. The GWR station in Gloucester was centrally located in the city. The stations at Newent and at Dymock were well situated in relation to the communities which they served, and were capable of dealing with many different types of traffic (discussed in Chapter 2). The station at Ledbury was at some distance from the town centre, but this disadvantage was offset by the opening in late November, 1928, of Ledbury Town Halt which was close to the main part of the town; and at times a bus service between the main station and the town centre was also available. Further attempts to stimulate local traffic were made by the provision in 1937 of halts at Greenway and at Four Oaks, and in 1938 of a halt at Malswick. (These halts are described in Chapter 3.)

From 1940 onwards many of the passenger services were provided by GWR diesel railcars. These brought operating economies, as they were cheaper to run than steam trains, and no fireman was required; and they perhaps had an air of modernity which attracted custom. The railcars were also well suited to the nature of the traffic along the line: unlike buses, they had ample space for parcels and perambulators; and they were capable of hauling an extra passenger coach if the demand arose.

During the Second World War the modal shift from rail to road was retarded by the petrol shortages which affected bus operators and the previously increasing number of car owners. The ending of hostilities in 1945 led to a partial return to pre-war trends, although the further transfer of traffic to the roads was hampered by the maintenance of petrol rationing, which continued until 1950 (and which was briefly re-introduced at the time of the Suez Crisis in 1956).

The rise of road transport was, however, only retarded and not reversed by these difficulties, and during the decade of the 1950s numerous passenger services throughout Britain were withdrawn. The nationalisation of most of the country's railways had taken effect on 1st January, 1948, when almost all of the assets of 'The Big Four' railway companies were taken into public ownership. The newly formed Western Region of British Railways proved to be particularly assiduous in its examination of its secondary and branch lines and of its wayside stations, many of which did not survive this scrutiny and which accordingly closed during the 1950s and early 1960s (thus anticipating the recommendations made in Dr Richard Beeching's 1963 report *The Reshaping of British Railways*). After more than seven decades of operation, the Gloucester to Ledbury line did not escape the closure programme. The passenger service was drastically reduced in the autumn of 1958, and complete withdrawal followed on 13th July, 1959. There being no Sunday service, the last passenger trains over the line ran on Saturday, 11th July, in that year.

From 13th July, 1959, the section of line between Ledbury and Dymock was closed completely. Stop-blocks were placed on the two running lines just to the south of the overbridge at the north end of Dymock station. Goods trains between Gloucester and Dymock

HISTORICAL OUTLINE

Ledbury Town Halt, after it was fitted with electric lighting. *Lens of Sutton Association*

Discussion at Newent, as railcar W19W pauses with a service from Gloucester to Ledbury on 8th July, 1959, shortly before closure of the line. *Author*

continued to run until and including Saturday 30th May, 1964, when they too were withdrawn, and the line from Over Junction to Newent and Dymock was finally closed.

Gradients

Because the railway largely replaced the canal, the gradients mainly corresponded with changes in the canal levels, and hence were mostly unremarkable, the major exceptions being the differing routes followed by the canal and by the railway between Newent and Dymock, and the climb into Ledbury town and to the junction with the Worcester and Hereford main line west of Ledbury station.

Upon leaving Over Junction the line to Ledbury rose slightly at 1 in 594 to attain the level of the canal as it traversed the plain of the River Leadon, and was then level for a distance of almost three miles before again rising slightly at 1 in 1452 at the approach to Barber's Bridge and at 1 in 380 just before and then through the station itself. This was succeeded by almost a mile of level track and then by rises at 1 in 230 and 1 in 333, a short level section, a further rise at 1 in 141, and lastly by a descent of 1 in 415 through Newent station.

Approximately half a mile beyond Newent station the line began to rise steeply at a gradient of 1 in 80 for nearly two miles to reach the high ground which Josiah Clowes had warned the canal proprietors to avoid. Their decision had necessitated the construction of the Oxenhall Tunnel (2,192 yards) beneath the hills. Having reached the summit, the railway had a brief level section in a deep cutting, the site of the future Four Oaks Halt. The line then descended at the same steep inclination of 1 in 80 for nearly one and a half miles before easing to 1 in 521 south of Dymock station. After the station the line then resumed its descent, this time briefly at 1 in 100. By now the railway had rejoined the line of the canal and of the river, and consequently gradients again became moderate: 1 in 330 up; very short level and downward intermissions; and then 1 in 520 up and 1 in 277 up to reach the pre-existing canal basin south of Ledbury. Finally there were two steep upward pitches, one at 1 in 72 as far as Ledbury Town Halt (corresponding to the five locks which raised the level of the canal to that of Ledbury town), and then another at 1 in 64 to enable the railway to attain the level of the Worcester and Hereford main line at Ledbury Junction.

HISTORICAL OUTLINE

The deep cutting at the summit of the line, between Newent and Dymock. Four Oaks Halt is visible on the right hand side of the track. *Author's collection*

Stations

The three intermediate stations (as distinct from the four halts) along the line were graced with buildings designed by the architect William Clarke, who was responsible for numerous station buildings situated on the Great Western system: some of these were built for independent railways which were (sooner or later) absorbed by the GWR.

Unfortunately little biographical or personal information about William Clarke is available, but it seems clear that he had a varied and successful career as an engineer, culminating in his appointment as Assistant Chief Engineer of the London and North Western Railway. He had previously served as the Resident Engineer of the Lahore Division

28 FROM GLOUCESTER TO LEDBURY: THE DAFFODIL LINE

Dymock looking south, taken from the minor road bridge at the north end of the station.
John Alsop

Passengers hasten to join the railcar W19W before it leaves Newent for Gloucester.
Lens of Sutton Association

of the Punjab Railway from 1859 to 1862, and upon his return to England was associated with various independent railways, most of which were in due course acquired by the Great Western Railway.

William Clarke designed a standard station building, which, with minor variations, was instanced in at least 24 locations, mostly in the West Midlands and in the west of England. The station buildings at Barber's Bridge, Newent and Dymock all followed his standard design. This was, essentially, for a single storey building of rectangular plan, constructed of brick and stone, and having distinctive, accented, stone quoins at the corners and window surrounds. The building had a simple pitched roof, which was surmounted by tall, ornate, and heavily built chimneys, again with stone quoins. There was a wide canopy extending the whole width of the platform, and inclined slightly downwards to the main building, so that rainwater could flow to a valley. This canopy was supported by several large brackets attached to the platform side main wall of the building, and there were no supporting posts on the platform itself. At each station between Gloucester and Ledbury, the spandrels (the space contained within the arch of the brackets supporting the canopy) included some decorative ironwork, fashioned with shields bearing the initials 'N. R.' (Newent Railway). The design of the supports and canopy meant that the free and covered flow of passengers, luggage and parcels was greatly facilitated and unobstructed. The resultant appearance of Clarke's stations was at once simple and practical, distinctive and pleasing.

Halts

None of the four halts along the line was provided with the 'pagoda' design of shelters which are often, and correctly, associated with the 'halts' and 'platforms' of the Great Western Railway. The first halt to be opened, by 28th November, 1928, was Ledbury Town Halt, and here a simple hut of corrugated-iron was provided for the accommodation of waiting passengers.

The other three halts – Greenway (1st April, 1937), Four Oaks (16th October, 1937) and Malswick (1st February, 1938) – were opened almost a decade later and within the space of one year. Each of them was adorned with a wooden shelter, further enhanced by a small valanced canopy, that at Greenway being slightly narrower than those at Four Oaks and at Malswick.

An auto-train bound for Ledbury stands at Dymock down platform, The engine is 'sandwiched' between two auto-trailers. *Colin Maggs*

Railcar No. W19W which was regularly used on the line, is seen here in open country near Barber's Bridge on 11 July, 1959. Note the well maintained track and the neat post and wire fences. *Hugh Ballantyne/Rail Photoprints*

Chapter Two

Operating The Line

Passenger Train Services

During the almost 74 years of operation, the passenger train service remained broadly the same. For most of this time it consisted of five trains, on weekdays only, in each direction between Gloucester and Ledbury. In the early years, certain, but not all, trains travelling from Gloucester continued to Malvern, and in some cases to Worcester (Shrub Hill), and vice versa. In 1902, for example, the first down train of the day departed from Gloucester at 7.23 am, and arrived at Ledbury at 8.03. Here it connected with the 8.22 (8.00 ex-Hereford) train to Worcester (Foregate Street) and Birmingham (Snow Hill), and with the 8.30 (7.35 ex-Worcester (Shrub Hill)) to Hereford. While the platforms were needed for these two main line trains, the branch train would wait in the stabling siding at the north end of the down platform, and then, after the departure of the Worcester to Hereford train at 8.30, run into the down platform to be ready to leave at 8.35 for Gloucester, where it arrived at 9.15.

At 10.38 am the train departed from Gloucester, arrived at Ledbury at 11.18 and one minute later, at 11.19, left to call at all stations to Worcester (Shrub Hill), where it arrived at 12.15 pm. Meanwhile a train which had left Worcester (Shrub Hill) at 9.35 am stopped at all stations to Ledbury, where it called from 10.27 to 10.30 before continuing to Gloucester, reached at 11.20. At Newent this train crossed with the 10.38 from Gloucester, the two trains departing from Newent at 10.59 and at 10.58 respectively.

The train which had originated at Worcester could now form the 12.30 pm service from Gloucester to Ledbury, where it terminated at

GLOUCESTER, LEDBURY, MALVERN AND WORCESTER.

[Timetable for January to April 1902 showing train times between Gloucester, Ledbury, Malvern and Worcester in both directions, Week Days only]

K—Change at Worcester (Foregate Street).

ROUTES FOR RETURN TICKETS.—For particulars, see page 166.

This timetable for January to April, 1902, shows the five passenger trains each direction daily service which applied throughout most of the history of the line

1.10, and where connexions were available by the 1.26 pm (12.50 ex-Hereford) train to Worcester (Shrub Hill) and by the 1.40 (12.55 ex-Worcester (Shrub Hill)) train to Hereford. The branch train could then leave Ledbury at 1.48 for Gloucester, where it arrived at 2.25.

The next working was the 3.25 pm train from Gloucester. It called at Ledbury from 4.11 to 4.15 and then ran semi-fast to Worcester (Shrub Hill), arriving there at 5.10 pm. From Worcester (Shrub Hill) a train had departed at 3.30, and, stopping at all stations, called at Ledbury from 4.23 to 4.25, and then departed for Gloucester, arriving there at 5.08.

There was now a short lull before the ex-Worcester train left Gloucester at 6.50 pm and ran to Ledbury, where it paused from 7.40 to 7.45, and then continued to Great Malvern, where it terminated at 8.05. From Great Malvern it left at 8.25 for its return journey to Gloucester, calling at Malvern Wells, Colwall, Ledbury (8.49 to 8.55), Dymock, Newent and Barber's Bridge, and arriving at the cathedral city at 9.35 pm. This was the last train of the day.

In 1907 the GWR made a small and slightly amusing concession. The distance from Paddington to Newent via Swindon and Gloucester was 124 miles whilst that via Worcester and Ledbury was 145 miles. The longer Worcester route did, however, make possible a later departure from Paddington, and the GWR ruled that holders of return tickets between Newent and Paddington via Gloucester could return via Worcester upon payment of half the difference between the respective fares (the resultant excess fares were: 1st class 1s. 4d.; 2nd class 0s. 11d.; and 3rd class 0s. 9d.).

This five train pattern of services continued largely unchanged up to and including the first part of the First World War. The same was true of passenger services in Britain generally. In the first part of the First World War, the majority of passenger trains in Britain continued to run normally, but in late 1916 services throughout the country were liable to reduction or to curtailment. In particular, this simplification meant that many through trains were discontinued, and from the end of the year trains from Gloucester proceeded no further than Ledbury. Passengers for stations beyond Ledbury now needed to change there, and passengers wishing to travel from Gloucester to Worcester could use the direct service provided by the Midland Railway. The through service from Gloucester via Ledbury to Malvern and to Worcester was never restored.

The Gloucester to Ledbury line now became one of mainly local importance, one which in France would be called *une voie ferrée d'intérêt locale*. At various times, and at the southern end of the line, some trains began or ended their journeys at Cheltenham Spa (St James's), 7½ miles from Gloucester Central. The five services in each direction could normally be provided by a single train. There was no regular service on

Sundays, although from time to time Sunday excursion trains were run, particularly during the spring when the wild daffodils of the beautiful 'Golden Triangle' (the area formed between Newent, Dymock and Kempley) were in bloom. These expanses of wild flowers lent to the line its soubriquet of 'The Daffodil Line'.

In 1932 there continued to be five trains in each direction along the branch on weekdays. The time of the first train from Gloucester had been advanced to 06.40 am, and subsequent departures were at 9.40, 12.35 pm, 4.05, and 7.00. Return trains left Ledbury at 8.15 am, 10.50, 1.50 pm, 5.20, and 8.25 (some of these times being almost unchanged from those which had obtained in 1902). The standard 1902 timing of 40 minutes for the journey between Gloucester and Ledbury had by 1932 been extended to 42 – 48 minutes, an increase which may be partly explained by the additional time needed to call at the new Ledbury Town Halt, situated at 59 chains from the junction.

By the summer of 1938, further halts had been added at Greenway (on 1st April, 1937), at Four Oaks (on 16th October, 1937) and at Malswick (on 1st February, 1938). Departures from Gloucester were now at 6.40 am, 9.39, 12.30 pm, 4.05, 6.40 on Mondays to Fridays, 7.15 (Saturdays only) and 9.50

The timetable for the summer of 1932. By this time Ledbury Town Halt has come into use.

From *Bradshaw's Guide* for July 1938. In addition to the standard five trains daily in each direction, there is a late evening return working on Saturdays only. By now all four halts listed below the table itself had come into operation.

GLOUCESTER AND LEDBURY (Third class only, limited accommodation. Week Days only.)

		a.m.	a.m.		p.m.	p.m.		p.m.				a.m.	a.m. p.m.		p.m.
Gloucester	dep.	6 40	9 55	...	2 55	6 25	...	7 35	...	Ledbury	dep.	7 30	11 30	5 10	8 55
Barber's Bridge	,,	6 52	10 7	...	4 7	6 36	...	7 47	...	Ledbury Town Halt	,,	7 52	11 37	5 12	8 57
Malswick Halt	,,	6 59	10 14	...	4 14	6 40	...	7 54	...	Greenway Halt	,,	8 0	11 45	5 18	9 5
Newent	,,	7 4	10 20	...	4 19	6 46	...	8 0	...	Dymock	,,	8 6	11 50	5 23	9 9
Four Oaks Halt	,,	7 9	10 26	...	4 25	6 57	...	8 5	...	Four Oaks Halt	,,	8 10	11 54	5 27	9 14
Dymock	,,	7 15	10 32	...	4 31	7 3	...	8 12	...	Newent	,,	8 15	12 1	5 33	9 20
Greenway Halt	,,	7 19	10 37	...	4 35	7 6	...	8 16	...	Malswick Halt	,,	8 18	12 4	5 36	...
Ledbury Town Halt	,,	7 27	10 45	...	4 43	7 15	...	8 25	...	Barber's Bridge	,,	8 26	12 13	5 45	9 30
Ledbury	arr.	7 30	10 50	...	4 46	7 18	...	8 35	...	Gloucester	arr.	8 36	12 25	5 58	9 45

S—For St. Davids.   Q—Saturdays excepted.   X—Third class only (limited accommodation).
L—Landore (Low Level).   Q—Pilning (Low Level).   S—Saturdays only.   †—Departure time.
U—Calls at Llanelly at 6.5 a.m. and Swansea (High St.) 6.50 a.m. to set down.   ¶—About 300 yards to Lydney Junction Station (S. & W. Line).

The timetable for October, 1945, still reflecting the wartime reduction of the service to four trains daily. The last train of the day from Ledbury does not call at Malswick Halt, an arrangement which mostly continued until the end of all passenger services in 1959.

GLOUCESTER AND LEDBURY (Third class only, limited accommodation. Week Days only.)

		a.m.	a.m.		p.m.	p.m.	p.m.					a.m.	a.m.		p.m.	p.m.	p.m.
Gloucester	dep.	6 42	9 20	...	12 5	...	4 5	6 25	7 35	Ledbury	dep.	7 55	...	10 42	...	1 30	5 30	8 5
Barber's Bridge	,,	6 54	9 32	...	12 17	...	4 17	6 35	7 52	Ledbury Town Halt	,,	7 57	...	10 44	...	1 32	5 12	8 8
Malswick Halt	,,	7 0	9 39	...	12 24	...	4 24	6 50	7 55	Greenway Halt	,,	8 5	...	10 52	...	1 40	5 20	9 3
Newent	,,	7 5	9 44	...	12 28	...	4 29	6 56	8 8	Dymock	,,	8 9	...	10 57	...	1 44	5 24	9 8
Four Oaks Halt	,,	7 10	9 50	...	12 34	...	4 35	6 57	8 5	Four Oaks Halt	,,	8 14	...	11 1	...	1 49	5 28	9 14
Dymock	,,	7 17	9 53	...	12 36	...	4 41	6 3	8 11	Newent	,,	8 18	...	11 6	...	1 53	5 35	9 20
Greenway Halt	,,	7 20	9 59	...	12 46	...	4 45	7 6	8 16	Malswick Halt	,,	8 22	...	11 10	...	1 57	5 39	...
Ledbury Town Halt	,,	7 29	10 7	...	12 52	...	4 53	7 15	8 25	Barber's Bridge	,,	8 30	...	11 19	...	2 5	5 45	9 30
Ledbury	arr.	7 32	10 10	...	12 55	...	4 57	7 18	8 27	Gloucester	arr.	8 42	...	11 34	...	2 17	6 2	9 45

S—For St. David's.   Q—Saturdays excepted.   L—Change at Stapleton   ¶—Third class only.   *—Llanelly arrive 9.45 p.m.
Road.   P—Cheltenham Spa (Malvern Road).   Q—Pilning (Low Level).   †—Clarbeston Road arrive 10.11 p.m.
S—Saturdays only.   X—Third class only (limited accommodation).   ‡—Arrival Time.
Z—Calls at Badminton at 5.44 p.m.   ¶—About 300 yards to Lydney Junction Station (S. & W. Line).

The timetable for October, 1947. By this time the standard service of five trains each way has been restored.

Table 108   GLOUCESTER and LEDBURY

		Week Days only								Week Days only			
Miles		a.m. X	a.m. S	a.m.	a.m. X X	p.m. X		Miles		a.m. X	a.m.	p.m. X	p.m. X X
—	185 London (Pad.) dep	1 0	..	..	7 20 8 22 12 15	..		—	185 Worcester (S.H.) dep	..	9 50	12 5	3 25 7 30
—	Gloucester Central dep	6 42	9 20	..	12 12 4 5 6 25	..		—	185 Great Malvern.. ,,	..	10 20	12 42	3 56 7 50
5½	Barber's Bridge.. ,,	6 54	9 31	..	12 24 4 17 6 36	..		—	165 Hereford ,,	7 20	7 55	12 40	4 48 7 43
8¾	Malswick Halt .. ,,	7 0	9 39	..	12 32 4 24 6 45	..		—	Ledbury dep	7 30	10 47	1 20	5 10 8 5
10	Newent ,,	7 5	9 43	..	12 37 4 26 6 50	..		2½	Ledbury Town Halt.. ,,	8 5	10 55	1 26	5 21 8 22
12	Four Oaks Halt ,,	7 10	9 48	..	12 42 4 34 6 56	..		3½	Greenway Halt ,,	8 10	10 55	1 40	5 21 8 22
13¾	Dymock ,,	7 17	9 55	..	12 49 4 41 7 2	..		7	Dymock ,,	8 14	11 1	1 43	5 24 8 27
15½	Greenway Halt ,,	7 21	9 59	..	12 54 4 45 7 6	..		7¼	Four Oaks Halt ,,	8 21½	11 5	1 45	5 40 8 46
19	Ledbury ,, arr	7 32	10 10	..	1 6 4 57 7 19	..		9	Newent ,,	8 22¼	11 10	1 50	5 46 8 50
—	185 Hereford arr	8 27 11 8	..	..	2 4 6 7 36	..		10¼	Malswick Halt ,,	8 22 1 14	1 54	5 53 8 53	
—	16 Great Malvern.. ,,	8 9 11 30	..	..	1 6 5 37 15	..		13½	Barber's Bridge ,,	8 30 1 22			5 58 9 17
54	165 Worcester (S.H.) ,,	8 28 11 52	..	..	1 52 5 58 9 1	..		19	Gloucester Central arr	8 42 11 36	2 15	6 10 9 17	
—									—	185 London (Pad.) arr	12 55 3 20	6 40 10 50	4d 25	

a a.m.   B Refreshment Car to Kingham.   Change at Kingham and Cheltenham Spa (Malvern Road).   F Forgate
Street.   B Refreshment Car from Swindon.   R Refreshment Car.   r Refreshment Car to Swindon.
X Third class only, limited accommodation.   A Third class only

The British Railways (Western Region) timetable for 20th September, 1954 to 12th June, 1955. The 'X' ('Third class only, limited accommodation') at the head of some columns indicates that the service is provided by a diesel railcar. The '3' at the head of the 9.20 am and 12:15 pm trains from Gloucester and their returns at 10.47 am and 1.26 pm from Ledbury almost certainly means that these services are provided by a steam train while the diesel railcar is making a round trip to Birmingham.

(Saturdays only). Return trains left Ledbury at 8.15 am, 10.50, 1.50 pm, 5.22, 8.28, and at 10.55 pm (Saturdays only: this last train did not call at Barber's Bridge). The late trains on Saturdays presumably catered for people wishing to spend an evening in Gloucester. The journey times were now between 47 and 56 minutes, and probably reflected the stops made at the four halts, the provision of which is evidence of an intention to increase traffic; and possibly pathing times (extra time added to a train's schedule to take account of conflicts with other trains) at Over Junction.

During the Second World War there was some reduction in the number of trains run, and by the end of 1945 there were only four trains running daily between Gloucester and Ledbury: departures from Gloucester were at 6.40 am, 9.55, 3.55 pm, and on Mondays to Fridays at 6.25 pm and on Saturdays at 7.35 pm; and from Ledbury at 7.50 am, 11.35, 5.10 pm and 8.55 (there were no differences between the return Monday to Friday and Saturday services from Ledbury). There were thus long intervals in the midday service from Gloucester and in the afternoon service from Ledbury. Some journey times were slightly increased. It is interesting that the quite late evening train from Ledbury still ran, and that it continued to omit Malswick Halt. Briefly in 1949 and 1950 there was an exception to this practice and it was possible to arrange, by request, for the evening up train – at that time the 8.25 pm departure from Ledbury – to call at Malswick on Mondays to Fridays. Intending passengers needed to give notice to the station master at Newent not later than 4.30 pm on the day of travel. Curiously, the working timetable emphasised that this facility was not advertised, and so presumably, it was only people having inside information who could avail themselves of the opportunity to alight from, or to board, the evening train at Malswick. From 1940 onwards many of the passenger services were worked by GWR diesel railcars, although some steam hauled trains and auto-fitted trains continued to run over the line.

By the end of 1947, the five trains daily service had been restored, with timings generally similar to those of the pre-war period. The autumn 1954 timetable showed some trains under the note 'X' which meant 'Third class only: limited accommodation', probably indicating that the service was provided by a diesel railcar. Other trains were adorned with the figure '3', denoting that the trains were 'Third class only', but suggesting that – there being no warning about limited accommodation – these particular trains were steam hauled and probably consisted of at least two coaches.

This pattern of service continued through the mid-1950s to be largely unchanged, with only minor alterations. The diesel railcars which

254   No. 7

GLOUCESTER AND LEDBURY.

Single Line.—Over Junction to Ledbury.—worked by Electric Train Staff. Over Junction to Ledbury Station. Crossing and Staff Stations.—Over Junction, Newent, Dymock, Ledbury Station. The Siding Points at Barber's Bridge are locked by a key fixed in the end of the Train Staff.
When Dymock Signal Box switched out, through Electric Token instruments between Newent and Ledbury brought into working.

WEEK DAYS ONLY

DOWN TRAINS

Distance from Gloster.	Mile Post Mileage from Over Jct.	STATIONS.	Gradient.	Point-to-Point Times.	Allow for Stop.	Allow for Start.	B Diesel arr.	B Diesel dep.	B Diesel arr.	B Diesel dep.	K Freight arr.	K Freight dep.	B Diesel arr.	B Diesel dep.	B Diesel S O arr.	B Diesel S O dep.
M. C.				Mins.	Mins.	Mins.	a.m.	a.m.	a.m.	a.m.	a.m.	a.m.	p.m.	p.m.	p.m.	p.m.
— —	—	GLOUCESTER	—	—	—	—	—	6 35	—	9 20	—	9 50	—	4 5	—	8 25
1 39	—	Over Junction	95 F.	4	—	1	6 42	C₈ 45S	C₈ 28S	9 32	R	10 21	C₁₂ 12 12	4 17	C₁₂ 28S	
5 37	3 78	Barber's Bridge	394 R.	11	—	1	6 54	6 55	9 39½	9 44	10 32	10 42	12 21	4 24½	6 36	6 38
7 47	5 7	Malswick Halt	230 R.	—	—	—	7 1	7 2	9 45½	9 46½	—	—	12 24½	4 27½	6 45	6 46
8 73	6 10	Newent	330 R.	12	1	1	7 4½	7 11	9 54	9 59½	10 57	X12 0	12 26	4 31	6 51	6 57
12 66	10 37	Four Oaks Halt	80 R.	—	—	—	7 16	7 17	9 54	9 59½	—	—	12 36	4 34½	6 57	9 52
13 66	12 27	Dymock	100 F.	—	—	—	7 21	7 22	9 54½	9 59½	12 14X	•12 58	12 37	4 42	9 59½	9 58
15 20	13 66	Greenway Halt	72 R.	4	2	—	7 26	7 27	—	—	—	—	12 41	4 46	—	9 70
18 70	16 61	Ledbury Town Halt	64 F.	—	—	—	7 29	7 30	—	—	—	—	12 49	4 54	7 15	7 10
		LEDBURY		14	—	1	7 32	—	10 10	—	1 10	—	12 52	4 57	7 18	—

UP TRAINS

		STATIONS.	Ruling Gradient.	Point-to-Point Times.	Allow for Stop.	Allow for Start.	B Diesel arr.	B Diesel dep.	B Diesel arr.	B Diesel dep.	K Freight arr.	K Freight dep.	B Diesel arr.	B Diesel dep.	B Diesel D arr.	B Diesel S O D dep.
				Mins.	Mins.	Mins.	a.m.	a.m.	a.m.	a.m.	p.m.	p.m.	p.m.	p.m.	p.m.	p.m.
		LEDBURY		—	—	—	—	7 55	—	10 42	—	1 30	—	5 24	—	8 25
		Ledbury Town Halt	64 F.	—	—	—	8 0	8 1	10 45	10 51	—	1 37	—	5 31	8 27½	8 36
		Greenway Halt	72 F.	—	—	—	8 4½	8 5	10 51½	10 52	—	1 40	5 32	5 35	8 36	8 40
		Dymock	80 R.	12	1	1	8 8½	8 9½	10 57	11 1	1 39½	1 49	5 35	5 36	8 40	10 29S
		Four Oaks Halt	90 F.	10	—	1	8 13	8 18	11 5½	11 11	1 43½	1 54	5 41	5 43½	8 46	10 34
		Newent	330 F.	—	—	—	8 22	8 29	11 18½	11 25	1 57	2 5	5 48½	5 49	8 50	10 41
		Malswick Halt	230 F.	11	1	1	8 29½	8 30½	11 25S	11 26S	2 4	2 5	—	5 51½	—A	10 42
		Barber's Bridge	594 F.	11	1	1	8 36	8 37	—	—	2 12	2 51	5 46	5 55	8 59	10 46
		Over Junction	95 R.	11	—	1	8 42	—	11 36	—	2 17	—	5 58½	C₈ 7S	C₁₀ 9 13	C₁₁ 15S
		GLOUCESTER		7	—	—	—	—	—	—	—	—	6 10	—	—	17 10S

Working of Heavy Engines over Newent Branch.—Engines of the 2-8-0 "E," Austerity and 2-8-0 "K" Types can run between Over Junction and Ledbury at a speed not exceeding 25 miles per hour subject to all Service restrictions. 4-6-0 ENGINES, 78XX "MANOR CLASS" OVER JUNCTION TO LEDBURY.—May work over the above Section at a speed not exceeding 25 miles per hour subject to existing prohibitions and restrictions. Newent.—Connection in Up Main Line between platforms calling at Back Siding off Goods Shed Rd. Speed not to exceed 4 m.p.h. Dymock.—Connection in Down Line between platforms leading to Goods Shed. Through connection being set down on notice being given by 4.30 p.m. to the Station Master at Newent, who will advise all concerned. NOT ADVERTISED. D.—Guard to extinguish lights where necessary. M.—For continuation see page 176.
R.—Starts from Old Yard. Docks Branch Sidings arr. 10.3 a.m. dep. 10.18 a.m. Z—Treated as a Halt. ‡—Advertised 12.0 noon.

The summer 1950 Working Timetable (which sets out (in note A) the slightly strange arrangement for the Malswick stop). Note that the Malswick concession did not apply on Saturdays!

Courtesy Michael Clemens

Newent, from just east of the station, and looking towards Dymock, 9th July, 1919. There was no overbridge at Newent, nor at any other station on the line, and passengers who needed to change platforms had to do so via the boards in the foreground.

John Alsop

Malswick Halt was accessed by these stairs from the road. The roof of the platform shelter can just be seen between the sign and the lamp post. Photographed shortly after closure of the line to passenger traffic, brambles are already taking over the climb.

Michael Clemens

usually provided the passenger service between Gloucester and Ledbury were stabled at Cheltenham. Here is a typical programme, namely that of the winter of 1957/1958. In the morning, a railcar departed from Cheltenham (St James's) station at 06.05 am, and, calling at Churchdown at 06.13, ran down to Gloucester Central, arriving there at 06.21. Here it waited to form the 06.42 to Ledbury, reached at 07.32, in good time for a connecting train, departing at 07.52, to Worcester, Kidderminster, Stourbridge Junction and Birmingham (Snow Hill). The railcar left Ledbury at 07.55 and arrived back at Gloucester at 08.42.

Subsequent departures from Gloucester were at 09.20 am, 12.10 pm, 4.08, and 6.24; and from Ledbury at 10.47 am, 1.25 pm, 5.25, and 8.22. It would seem that the 1.25 pm service from Ledbury continued from Gloucester Central to Cheltenham Spa (St James'), where it arrived at 2.46 pm, and from where it returned at 3.20 pm for Gloucester Central to become the 4.08 service to Ledbury. This extension (in both directions) to and from Cheltenham Spa was available to passengers, and perhaps also afforded the opportunity for a light servicing of the railcar at Cheltenham Spa. For most of the day, Barber's Bridge was treated as a halt, with the station being staffed for only a brief period during the middle of the day.

Trains called at all stations and halts along the line, except that, a little curiously, the 10.47 am and 8.22 pm trains from Ledbury did not stop at Malswick Halt. The guard of the last train of the day, the 8.22 pm from Ledbury, extinguished the lights at the halts, and for this purpose both Dymock and Barber's Bridge stations were additionally so classified.

On 6th September, 1930, the signal box at Dymock (nearly 5¼ miles from Ledbury) had been provided with a switch, which, when operated, enabled a 'long section' (of 9 miles) between Ledbury and Newent to be instituted. This facility left the branch to be controlled by three signal boxes, namely Ledbury, Newent, and Over Junction. This simplification meant, for example, that in the winter timetable for 1957 to 1958 the box at Dymock did not need to be opened until 10.35 am (in time for the 10.47 train from Ledbury to Gloucester, which called at Dymock at 11.00), unless the box was required to be opened earlier to enable the 09.20 train from Gloucester to attach or detach vehicles. It could be closed at 5. 55 pm, after the passage, at 5.39 pm, of the 5.25 pm train from Ledbury to Gloucester and its arrival at the next block post, Newent, at 5.50. Once this train was belled 'out of section', the Dymock signalman could close his box, and probably the station. In this way, the signal box at Dymock could be staffed by means of a single shift.

During the mid 1950s, a semi-fast, limited accommodation, mid-morning service was provided between Gloucester and Birmingham,

K94 WEEKDAYS

GLOUCESTER AND LEDBURY

SINGLE LINE—Over Junction to Ledbury—Worked by Electric Train Token, Over Junction to Newent, and Electric Train Staff Newent to Ledbury.

Mileage from Gloucester M	Mileage from Gloucester C	Mile Post Mileage from Over Jcn. M	Mile Post Mileage from Over Jcn. C	DOWN		Ruling Gradient 1 in	B 6.5 a.m. Cheltenham Diesel	N—Treated as a Halt	B Diesel	N—Treated as a Halt	B Diesel		B Diesel	N—Treated as a Halt on Sats.	B Diesel	Z—Treated as a Halt ‡—6.34 p.m.
							am		am		PM		PM		PM	
—	—	—	—	GLOUCESTER CEN. dep	1	—	6 42		9 20		12 10	..	4 8		6‡25	
1	39	—	—	Over Junction	2	95F	6 45		9 23		12 13	......	4 11		6 28	
5	37	3	78	Barber's Bridge arr	3	594R	..		..		..	..	..		..	
				dep	4		6N54		9N32		12 22	......	4N20		6 41 Z	
8	47	7	8	Malswick Halt	5	230R	7a 1½		9a39½		12 29½	..	4a27½		6 48½	
9	73	8	34	Newent arr	6	330R	7 4½		9 42½		12 32½	......	4X30½		6 51½	
				dep	7		7 6		9 44½		12 35	..	4 33		6 53	
12	6	10	47	Four Oaks Halt	8	80R	7 11½		9 50		12a40½	......	4a39		6a59	
13	66	12	27	Dymock arr	9	80F	7 16		9 54½		12 45	..	4 44		7 3½	
				dep	10		7 17½	N	9 55½	N	12 46	......	4 45½		7 5	
15	25	13	66	Greenway Halt	11	100F	7a21½		9a59½		12a50½	..	4 49½		7c 9	
18	20	16	61	Ledbury Town Halt	12	72R	7 29		10 7		12 58	......	4 57	N	7a16	Z
18	79	—	—	LEDBURY arr	13	64R	7 32	..	10 10	..	1 1	..	5 0	..	7 19	..

WEEKDAYS K95

LEDBURY AND GLOUCESTER

STAFF STATIONS—Over Junction, Newent, Dymock, Ledbury Station.
CROSSING STATIONS—Over Junction, Newent, Dymock.
The Siding Points at Barber's Bridge are locked by a key fixed in the end of the Train Staff.
When Dymock Signal Box switched out, through Electric Token instruments between Newent and Ledbury brought into working.

UP		Ruling Gradient 1 in		B Diesel	N—Treated as a Halt	B Diesel	B Diesel SUSPENDED SO	B Diesel	N—Treated as a Halt on Sats.	B Diesel	N—Treated as a Halt	B Diesel	N—Treated as a Halt Guard to extinguish lights where necessary			
				am		am	am	PM		PM		PM				
LEDBURY dep	1	—	..	7 55		10 47	10 47	1 26		5 25		8 22		..	..	..
Ledbury Town Halt	2	64F		7 57		10 49	10 49	1 28		5 27		8 24½		......	......	......
Greenway Halt	3	72F	..	8a 5		10a57	10a57	1a36		5a35		8 33		..	..	..
Dymock arr	4	100R	......	8 8½		......	......	......		5 38½		8N36½		......	......	......
dep	5		..	8 9½		11 1	11 1½	1 40½		5 39½		8 42		..	..	..
Four Oaks Halt	6	80R	......	8a14		11a 5½	11a 7	1a46		5a44		8 48		......	......	......
Newent arr	7	80F	..	8 18		11 9½	11 11	1 50		5 48		8 52		..	..	..
dep	8		......	8 19		11 11	11 13	1 52		5 51		8 53		......	......	......
Malswick Halt	9	330F	..	8a22½		..	11a16½	1a55½		5a54½		..		..	..	..
Barber's Bridge arr	10	230F	......	8N29½	N	11 20	11 23½	2N2½		6N1	N	9 2	N	......	......	......
dep	11		..	8 30½		11 20½	11 24½	2 3½		6 2½		9N3½		..	..	..
Over Junction	12	594F	......	8 39		11 29	11 33	2 12	N	6 13		9 12		......	......	......
GLOUCESTER CEN. arr	13	95R	..	8 42	..	11 32	11 36	2 15	..	6 16	..	9 15	..	..	..	..

The Working Timetable for passenger trains from 16th September, 1957, until 1st June, 1958. All services are now provided by diesel railcar (the daytime return journey to Birmingham by the car having been discontinued).

FROM GLOUCESTER TO LEDBURY: THE DAFFODIL LINE

Table 108   **GLOUCESTER and LEDBURY**
WEEK DAYS ONLY—(Second class only)

Miles		am	am	am	pm			am	pm	pm	pm
	105 London (Pad.) .. dep	..	9 5	11 45	2 15	Miles	164 Worcester (S.H.) .. dep	..	12F57	3 25	7 28
							164 Great Malvern ... „			1 10	3 56	7 48
—	Gloucester Central. dep	5 42	12 15	4 8	6 24		164 Hereford .. .. „	7 20		12 55	4 28	6 55
5½	Barber's Bridge	6 53	12 26	4 19	6 38	only			only			
8½	Malswick Halt	7 1	12 34	4 27	6 47		— Ledbury .... dep	7 55		1 30	5 25	8 22
10	Newent	7 5	12 40	4 32	6 52	days	½ Ledbury Town Halt . ..	7 57	days	1 32	5 27	8 24
12	Four Oaks Halt	7 11	12 45	4 38	6 58		3½ Greenway Halt	8 5		1 40	5 34	8 32
13½	Dymock	7 16	12 50	4 45	7 4	Saturdays	5½ Dymock	8 9	Saturdays	1 44	5 39	8 38
15½	Greenway Halt	7 21	12 55	4 49	7 8		7 Four Oaks Halt	8 13		1 48	5 44	8 46
18½	Ledbury Town Halt ..	7 28	1 3	4 56	7 16		9 Newent	8 16		1 53	5 50	8 52
19	Ledbury .... arr	7 32	1 6	5 0	7 19		10½ Malswick Halt	8 22		1 56	5 54	..
							13½ Barber's Bridge	8 30		2 4	6 2	9 2
32½	164 Hereford .. .. arr	8 37	1 46	6 10	8 30		19 Gloucester Central .. arr	8 42		2 17	6 16	9 15
26	164 Great Malvern ... „	8 10	1 32	5 59	8 47							
34½	164 Worcester (S.H.) .. „	8F26	1 52	6 10	9F 1		133 105 London (Pad.) .. arr	12 25		5 35	10 50	..

B Change at Kingham and Cheltenham Spa (Malvern Rd.) (Tables 173 and 110)   F Foregate Street   p pm

232

In the winter timetable for 15th Sepember 1958, to 14th June,1959, the service has been drastically reduced. From Mondays to Fridays there is now no departure from Gloucester between 6.42 am and 4.08 pm, and no return service from Ledbury between 7.55 am and 5.25 pm. Only on Saturdays is there a mid-day service, and even then the total number of trains in each direction is only four.

Table 108   **GLOUCESTER and LEDBURY**
WEEK DAYS ONLY—(Second class only)

Miles		am	am	pm	pm			am	pm	pm	pm
	105 London (Paddington) .. dep	..	7 30	11C45	2 15	Miles	164 Worcester (Foregate St) dep	..	12 47	3 30	7 32
							164 Great Malvern .... „			1 3	3 56	7 48
—	Gloucester Central .... dep	5 42	12 10	4 8	6 24		164 Hereford .. .. .. „	7 20		12 55	4 26	6 55
5½	Barber's Bridge	6 53	12 21	4 19	6 38	only			only			
8½	Malswick Halt	7 1	12 29	4 27	6 47		— Ledbury .... dep	7 55		1 30	5 25	8 22
10	Newent	7 5	12 35	4 32	6 52	days	½ Ledbury Town Halt	7 57	days	1 32	5 27	8 24
12	Four Oaks Halt	7 11	12 40	4 36	6 58		3½ Greenway Halt	8 5		1 40	5 34	8 32
13½	Dymock	7 16	12 50	4 45	7 4	Saturdays	5½ Dymock	8 9	Saturdays	1 44	5 39	8 38
15½	Greenway Halt	7 21	12 56	4 49	7 8		7 Four Oaks Halt	8 13		1 48	5 44	8 46
18½	Ledbury Town Halt	7 26	1 3	4 56	7 16		9 Newent	8 16		1 52	5 50	8 52
19	Ledbury .... arr	7 32	1 5	5 0	7 19		10½ Malswick Halt	8 22		1 56	5 54	..
							13½ Barber's Bridge..	8 33		2 4	6 2	9 2
32½	164 Hereford ...... arr	8 37	1 46	6 10	8 30		19 Gloucester Central .... arr	8 42		2 17	6 16	9 15
26	164 Great Malvern .... „	8 9	1 30	5 52	8A 5							
34½	164 Worcester (Foregate St) „	8 26	1 45	6 5	8A20		133 105 London (Paddington).. arr	12B25		5 40	10 10	..

A On Saturdays arr Great Malvern 8 47 pm, Worcester (Foregate St.) 9 1 pm   B On Saturdays arr 12 35 pm
C am. Via Kingham and Cheltenham Spa (Malvern Rd.) (Tables 173 & 110)   p pm

863

The final timetable, 15th June, 1959, to 13th September 1959. Very similar to that of the previous winter (15th September, 1958 to 14th June, 1959), this impoverished timetable was current when passenger services between Gloucester and Ledbury were withdrawn on Monday, 13th July, 1959, the last train, the 8.22 pm from Ledbury to Gloucester, having run on Saturday, 11th July, 1959.

with a return service in the early afternoon. The heading note in the public timetable was 'X', indicating that the train was third class only and of limited accommodation. This marking suggests (as in many other cases; and see above, page 35) that the service was operated by a diesel railcar.

For example, in the summer of 1955, the train left Gloucester (Central) at 9.55 am, and ran to Cheltenham Spa (Malvern Road), arriving there at 10.07. It departed a minute later, and, calling only at Broadway (10.32), continued to Stratford-upon-Avon, where it stopped briefly from 10.51 to 10.52. Then, with one stop, at Henley-in-Arden (11.06), it ran fast to Birmingham (Snow Hill), arriving there at 11.31 (11.33 on Saturdays). The return journey from Snow Hill began at 12.27 pm. This time there was no stop at Henley-in-Arden, and, with calls at Stratford-upon-Avon (1.08 to 1.10) and at Broadway (1.32), the railcar arrived at Cheltenham Spa (Malvern Road) at 1.58. Here it seems to have paused until 3.22, when it ran down to Gloucester (Central), reached at 3.36. To achieve a faster onward journey passengers from Birmingham and Stratford-upon-Avon could change at Malvern Road into the 2.12 (2.42 on Saturdays) train to Gloucester. It is possible that during the interval the railcar underwent a light servicing.

It is said that the railcar assigned to the services just described was 'borrowed' from its more usual duties on the Gloucester and Ledbury line. Such an arrangement is certainly compatible with the known facts of the case. The first round trip of the day (beginning at 6.42 am from Gloucester (Central)) was completed at 8.42 am, when the railcar arrived back in Gloucester. There it would be available to take up the 9.55 am working to Birmingham, and on its return to Gloucester at 3.36 pm it would be in ample time to form the 4.08 to Ledbury and return, and subsequently the 6.25 to Ledbury and return. So what of the interim on the Ledbury line? It would seem likely that, during the absence of the railcar on its journey to Birmingham and back, the 09.20 am train from Gloucester to Ledbury and return, and similarly the 12.12 pm to Ledbury and return, were worked by steam trains, either by an auto-train (locomotive and trailer) or by a locomotive with (probably) two third class coaches. Any such arrangements had probably ceased to apply by the autumn of 1957, by when the Gloucester to Birmingham and return semi-fast services had disappeared from the public timetable, and by when also the working timetable for the Gloucester and Ledbury line indicated that all passenger trains over the line consisted of a diesel railcar.

The five train service, with slight variations in timings, continued until the end, on 14th September, of the summer timetable for 1958. The autumn of 1958, however, brought some ominous changes. The 9.20 am train from Gloucester and the return 10.47 from Ledbury were discontinued altogether, and the mid-day trains now ran only on Saturdays, leaving Gloucester at 12.15 pm and returning from Ledbury at 1.30. This meant that on Mondays to Fridays there was a long gap of more than nine hours in each direction: there was no train from Gloucester between 6.42 am and 4.08 pm; and no train from Ledbury between 7.55 am and 5.25 pm. For practical purposes, much of the value of the line for passengers was lost. It is difficult to resist the inference that these changes were part of a deliberate plan to run down the line with a view to closure.

This reduced timetable persisted into the summer of 1959, and it was unsurprising that the service was withdrawn altogether on Monday, 13th July, 1959. There being no Sunday service, the last trains ran on Saturday, 11th July, with the very last working being the 8.22 pm from Ledbury, which was due to arrive at Gloucester at 9.15. This consisted of five coaches headed by engine No. 3203. So ended almost 74 years of passenger services over the line.

Goods traffic

Prior to the opening of the railway between Gloucester and Ledbury, the least indirect route for traffic between Gloucester and Worcester and the West Midlands had been via Ross and Hereford, from where the way lay via Ledbury and Malvern. By this route the journey from Gloucester to Ledbury was 43½ miles, to Worcester (Shrub Hill) 59¼ miles, and to Birmingham (Snow Hill) 92¾ miles. When, however, the direct line from Gloucester via Newent and Dymock came into use, the distance to Ledbury was reduced to 19 miles, and this saving of 24½ miles resulted in a journey of 34¾ miles from Gloucester to Worcester (Shrub Hill) and 68¼ miles to Birmingham (New Street). These shorter distances represented a notable improvement, although they were still appreciably longer than those of the Midland Railway (MR) route from Gloucester, from where Worcester (Shrub Hill) was 25¾ miles and Birmingham (New Street) 51¾ miles distant. Both routes suffered from operational handicaps: the GWR trains had to contend with the steeply graded single bore tunnels at Ledbury and at Colwall, and the Midland with the formidable Lickey Incline between Bromsgrove and Blackwell. These through routes, were, however, of great importance to both companies: Gloucester itself generated much traffic, and many goods came via the Gloucester and Berkeley Ship Canal, which ran from a junction with the River Severn at Sharpness into the extensive Gloucester Docks, where the GWR and the MR each had a complex system of wharves and sidings.

So far as the GWR was concerned, the situation was to be transformed by the opening of a new line from Cheltenham to Honeybourne completed in 1906. This was one of a number of well engineered 'cut-off' lines built by the GWR early in the 20th century to shorten journeys between various centres and also to enable higher speeds to be attained. In addition, the opening of the North Warwickshire line between Bearley (north of Stratford-upon-Avon) and Tyseley (south of Birmingham) contributed to a reduction in distances and journey times. The combined route was opened to goods traffic on 9th December, 1907, and to passenger trains on 1st July, 1908. The distance between Gloucester Central and Birmingham (Snow Hill) was now 61 miles, and was over an appreciably easier route. Traffic for Worcester and the West Midlands generally could use the new line as far as Honeybourne (20 miles from Goucester (Central)), and then join the former Oxford, Worcester and Wolverhampton line to reach Worcester (Shrub Hill), 38¾ miles from Gloucester (Central): this was slightly further than the route via Ledbury (34¾ miles), but was more easily negotiated.

Dymock, looking south: the goods shed can be seen behind the main station building with wagons in the sidings there. *John Alsop*

The availability of the new lines from Cheltenham to the Birmingham area led to the diversion of nearly all through goods trains away from the Gloucester and Ledbury direct line. This led in turn to the downgrading of the line, but was also to the relief of the operating department, and in particular to that of enginemen who no longer had to undergo the stifling and indeed dangerous conditions prevailing in the single line tunnels between Ledbury and Malvern. (Freight trains to and from Hereford and the 'North and West' line did of course continue to run over the Ledbury and Worcester section.) But from now on (1908), the goods trains which ran between Gloucester to Ledbury were for primarily local purposes.

There were, however, occasional exceptions. For example, in 1947 the Strangford viaduct near Fawley on the line between Gloucester and Hereford sustained severe flood damage, and in consequence the line between Ross and Hereford was severed. As a result, goods trains between the two cities were for several months diverted to the Gloucester and Ledbury line, with a reversal at Ledbury. Often these trains were heavier than those which were usual on the local line, and would require assistance for the steep ascents to Ledbury Town and to the junction. Such assistance was provided by the Ledbury banking engine, whose primary task was to assist up trains through Ledbury Tunnel. At other times, London Midland and Scottish Railway (LMSR) (and, later, British Railways (London Midland Region)) goods trains would use the Ledbury route if that company's direct line from Gloucester to Worcester (and beyond) was closed for engineering work.

We have already seen that the hoped for coal traffic originating in the Newent area did not materialise. Attempts were made to gain access to the coal measures, but even when these could be reached, the coal proved to be meagre in quantity and indifferent in quality. This was a great disappointment. Instead of conveying coal away from Newent, the canal and then the railway were in the event to be used to bring to the area the good quality coal which, in the absence in most of the countryside of gas and electricity, was required all the year round for domestic and for some small scale industrial purposes.

There were other forms of goods traffic. Agricultural lime was constantly needed to improve the fertile but heavy soils of the area served by the line. Machinery was also brought in by rail. Goods outwards included livestock (ranging from cattle to chicks), dairy products such as milk, cream and butter, and market garden produce, including vegetables and flowers. Most spectacular amongst the latter were the vast quantities of daffodils – which gave the line its soubriquet of 'the Daffodil Line' – and which for a period of nearly a month each spring were consigned to the London area (via Gloucester) and to the West Midlands (via Ledbury). Many of these flowers were wild daffodils (*Narcissus pseudonarcissus*) which grew in almost prodigal profusion within a 'Golden Triangle' lying between the villages of Newent, Dymock and Kempley. These beautiful flowers did not enjoy a long 'shelf life', and before despatch they were often tied with raffia into bunches and posies, so that once they reached their destinations there would be no delay in placing them on sale, or, often, in sending them to hospitals to cheer the patients. Conversely, many town dwellers used the train to visit the area and to buy flowers direct, thus creating a welcome source of income for agricultural workers in the district. The usual 'No Sunday Service' rule on the line was suspended for a few weeks in the spring in order to enable special trains from Gloucester and from further afield (occasionally from London) to visit the 'Golden Triangle' for excursionists to view the carpets of daffodils and perhaps to buy a posy to take home. From 1925 onwards residents in the 'Golden Triangle' would, upon an agreed date or weekend, gather daffodils to be

Wild daffodils beside the Daffodil Way, in an orchard near Allum's Farm, Dymock March 2022.
Philip Halling

At Newent a Gloucester bound train enters the up platform, upon which is the signal box.
John Alsop

sent to London hospitals; and the made up bunches and posies would then be conveyed and delivered free of charge by the GWR.

In rural areas the inter-war period saw the rapid rise of local goods transport services provided by independent carriers using their own motor vehicles. To counter this tendency, the GWR developed its own methods. From Dymock and Newent stations the GWR's own system of 'country lorries' came into operation, whereby goods could be delivered to and collected from surrounding villages and settlements by road, with the stations being railheads for such traffic. Barber's Bridge station was not itself a 'country lorry centre', but its surrounding area could be served by 'country lorries' from other centres, including Newent.

The three intermediate stations all handled the kinds of traffic typical of country stations in agricultural areas, and in addition each had its own speciality.

Barber's Bridge handled incoming coal and agricultural necessities and general goods, as well as livestock including horses, cattle, and chicks. Outbound traffic comprised dairy products and seasonal fruits and soft fruits and vegetables, especially tomatoes. An additional and unusual form of traffic was pitch, which was brought by lorries from Severn refineries to be loaded (with the aid of a concrete ramp) into wagons which were mainly destined for South Wales and for the north of England. A crane of three tons was provided.

Newent dealt with a similar range of coal, farming commodities, dairy and horticultural produce. A crane of five tons capacity was provided. Feedstuffs for animals were held in a store which was separate from the main goods shed. Newent's special traffic proceeded from the Lancaster Sawmills, situated to the south of the main part of the station. Coal merchants included Meates, Hyetts, and Fields. A service of collecting and delivering goods at Newent was provided by the firm of R. T. Smith, who operated cartage facilities in various parts of Gloucestershire, although later these were partly overtaken by the GWR 'country lorry' system just described.

Dymock's range of merchandise was broadly similar to that of Newent, and this station too had a special, dedicated, building to hold agricultural feedstuffs. As at Newent, a five ton crane was available to assist with loading and unloading. A distinctive form of traffic was cider, made by the firm of Henry Weston in the Herefordshire village of Much Marcle, which lay to the west of Dymock; and in addition there was the seasonal and large volume of daffodils already described above. In addition, the goods yard dealt with timber for pit props destined for use in South Wales collieries, and with large quantities of soft fruits, apples, and hops. In the autumn, substantial volumes of sugar beet were despatched, many of them being sent to Kidderminster for processing there.

No provision was made at the halts for handling goods traffic.

After complete closure on 13th July, 1959, of the section between Ledbury and Dymock, the line between Over Junction and Dymock was treated as a single section. At Newent the pointwork for the goods yard was controlled from a ground frame unlocked by the train staff for the new long section. The running loop points were worked by hand, and when not in use were normally clipped by a padlock, the key for which was held in the station master's office. Until at least 1962 one goods train ran daily in each direction between Gloucester and Dymock, but from the summer of 1963 onwards the service was reduced to Tuesdays, Thursdays and Saturdays only.

Motive power

Although most of the railway between Gloucester and the southern side of Ledbury followed the line of the canal and so was easily graded, there was the formidable 'gable' with upward and then downward gradients of 1 in 80 between Newent and Dymock as the railway crossed the *massif* beneath which the canal ran in the Oxenhall Tunnel. Accordingly some at least moderately powerful locomotives were required, especially for freight trains, which had to be hauled up to the summit and then controlled on the ensuing descent. There were also the steep gradients (1 in 72 and 1 in 64) needed to lift the line into Ledbury town and then to the junction with the Worcester and Hereford main line. The first section, at 1 in 72, corresponded to the five locks required to raise (by about 60 feet) the level of the canal into Ledbury itself; and the second ascent was required to bring the line to the level of the Worcester to Hereford line to the west of Ledbury station.

For these purposes some engines of the 4-4-0 wheel arrangement were commonly used, and included some members of the 3252 ('Duke') class. Engines of the 2301 class 0-6-0s ('Dean Goods') – for example, 2349 – were also found to be suitable, as later were 0-6-0 tender engines of the 2251 class, such as 3219 and 3203: the latter engine also worked the very last passenger service on the line. Tank engines used included members of the 2-6-2 4500 (quite frequently 4573) and 5500 classes. The 0-4-2 tank 4800 (later 1400) class was also represented, as such engines could also work auto-trains. Examples were 1401, 1424, 1427 and 1428. If for any reason an auto-trailer was not available, these 1400 tank locomotives could simply haul conventional carriages. Conversely, non auto-fitted locomotives could haul an auto-trailer. In either case, the engine could run round at the terminus. At various times 0-6-0 pannier tank engines, such as 8717, also appeared on the line. Engines of the 4-6-0 7800 'Manor' series were also permitted to work over the line (see below, on weight and speed restrictions), but it would seem that they did so only rarely.

From July 1940, many of the passenger services were provided by GWR diesel railcars. This interesting form of vehicle had originated in 1933, and by April 1936, a total of 17 such cars were in service. Essentially they were single railcars with streamlined bodies, and were popular with passengers – so much so that at times the services which they were diagrammed to work had to be provided with substitute steam trains with their greater carrying capacity! Entering service in 1937, railcar No. 18 marked a turning point: it was provided with an angular body and was fitted with buffers and drawgear enabling it to haul a trailing load, usually consisting of one, or occasionally two, conventional coaches. This design was also successful, and, with further modifications, 19 more railcars were completed between July 1940, and February 1942. It is understood that railcars Nos. 25 and 29 were at times assigned to the Gloucester and Ledbury service in the early years, and later No. 19 was for some years, especially after 1955, the regular railcar

OPERATING THE LINE

The north end of Gloucester station looking along the track towards Over Junction.
John Alsop

Auto fitted Collett 0-4-2 tank englne No. 1428 pauses at Newent's down platform with a train from Gloucester Note the arched rodding tunnel on the face of the up platform: through this tunnel formerly passed point rods and signal wires from the McKenzie and Holland signal box located on the platform. *Tony Harden collection*

on the line. At busy times it would haul a standard compartment coach to provide extra accommodation.

When the diesel railcar was not available for service, or possibly when it was required for other duties (above, pages 38 – 41), the passenger trains would usually consist of either an auto-trailer in the care of a Collett class 1400 0-4-2 tank locomotive (such as 1428) or of a conventional two or three coach train headed by a 2-6-2 tank locomotive (such as 4564 or 4573).

The GWR (and subsequently British Railways (Western Region)) depot at Gloucester was named Gloucester Horton Road, in order to differentiate it from the LMSR (and subsequently British Railways (London Midland Region)) depot known as Gloucester Barnwood. Horton Road shed (coded 85B) came within the domain of the Worcester District main shed (coded 85A), but for practical purposes was largely autonomous: it had a large allocation of locomotives and was also responsible for several sub-sheds, including that located at Cheltenham (Malvern Road). Other sub-sheds were situated at Lydney, Brimscombe, Cirencester and Tetbury. Gloucester Horton Road depot and its Cheltenham sub-shed together supplied the motive power for trains traversing the line between Gloucester and Ledbury. The diesel railcar was kept and serviced at Cheltenham (Malvern Road).

After the closure of the line to passenger services, goods trains continued to run between Gloucester and Dymock, and would seem to have been always steam hauled. The locomotives continued to be

The up starting signal at Newent is clear for railcar No. W19W to depart for Gloucester when staff discussion is complete. *Lens of Sutton Association*

provided by Gloucester Horton Road shed, and they included members of the 2251 class (for example, 2232, 2248, 2273 and 2291), and prairie tank engines 4564 and 5518. The latter engine was withdrawn in May 1964, shortly before the cessation of the goods service to Dymock, and was replaced by 5545, imported from Southall (Middlesex). Towards the time of final closure, British Railways standard 2-6-0 locomotives of the 78000 series were also used, and the last freight trip of all, made on Saturday, 30th May, 1964, was worked by engine 78001.

Presumably because of the steep gradients over the final, northern, section of the line, steam locomotives usually ran chimney first towards Ledbury, as this practice would help to ensure that the firebox remained covered by water. There being no water cranes for locomotives at intermediate points along the line, engine crews were careful to ensure that, before beginning journeys, their locomotives had sufficient water in their tender or tanks.

Weight and Speed Restrictions

The route colour of the line between Over Junction and Ledbury was 'dotted blue': this meant that the line could be used by engines of the 'uncoloured' and 'yellow' weight classifications, that is, by engines having an axle load of up to 16 long tons; and it could also be used by engines classifed as 'blue', that is, by engines having an axle load of up to 17 long tons and 12 cwt., provided that they did not exceed a speed of 25 mph or any other lower limits applicable to particular lengths along the line. Accordingly, engines of the 7800 'Manor' class (which had an axle load of 17 tons and 5 cwt.) were permitted over the line, but again were not allowed to exceed 25 mph at any point; and they had of course also to observe any localised lower speed limits, and in addition were not allowed to exceed 4 mph when traversing certain parts of the layouts at Newent and at Dymock.

A variety of speed restrictions applied throughout the line. Trains entering or leaving the line at the sharply curved Over Junction were required not to exceed 10 mph. Thereafter limits of between 30 mph and 45 mph applied as far as the eastern end of the Newent station layout, throughout which a limit of 15 mph had to be observed. A similar restriction was imposed upon the layout at Dymock station. Between these two places, however, a limit as high as 50 mph was set over the 'gable', with its approach gradients of 1 in 80 on either side of the summit. Naturally such a speed would be attained only by trains descending the inclines. North of Dymock, limits of between 30 mph and 45 mph again applied as far as the curves west of Ledbury, where trains were restricted to 15 mph while negotiating the junction with the Worcester to Hereford main line.

Ledbury Locomotive Depot

It is said that one or more of the tender engines used on the line between Gloucester and Ledbury were provided with specially modified tenders to enable them to use the short turntable which was installed at Ledbury, where the locomotive depot was of a rudimentary kind. This statement may, however, be not quite correct, since many GWR locomotives of the 4-4-0 wheel arrangement had small tenders to enable the engines to be turned on the short turntables which were common on various parts of the GWR system, especially west of Newton Abbot.

It would seem that no shed was ever provided at Ledbury: there was only a simple coaling stage and a water supply. These were needed especially for the banking tank engines which assisted trains through the steeply graded and small bore Ledbury Tunnel (1,323 yards). The 'Ledbury bankers' were also occasionally used to assist freight trains ascending the steep inclines (at 1 in 72 and 1 in 64) between the site of Ledbury Basin and Ledbury Junction. These engines did not, however, normally need to use the turntable, since the usual practice was for them to run bunker first when assisting trains through Ledbury Tunnel. In this way the chimney was at the rear end of the engine, and hence in the tunnel there were fewer fumes for the engine crews to inhale.

Signalling and Single Line Working

The signalling for the line between Gloucester and Ledbury was first provided by the firm of McKenzie and Holland of Worcester. In practice, signals were required at Barber's Bridge, until the layout was simplified there, at Newent, and at Dymock. The signalling at Over Junction and at Ledbury did of course require some modifications when the new line was opened, and these changes were made partly at the expense of the two local companies (which at the time of the opening of the line were still constitutionally independent of the GWR).

When the line opened, operation of the single line sections between Over Junction and Newent, and between Newent and Dymock was by the train staff, and then by the train staff-and-ticket, systems. Whereas the former method presupposed that trains would always run alternately from one end of a section to the other, the latter enabled two (or more) trains to be worked consecutively in one direction. The train staff for the Over Junction to Newent portion of the line was triangular in section and coloured yellow, whilst that for the Newent to Dymock portion was round in section and coloured blue. (It will be remembered that until 1916 there was double track between Dymock and Ledbury.)

In July 1898, these methods were replaced by the electric staff system. This system remained in regular use between Newent and Dymock and (from 1917) between Dymock and Ledbury until the closure of the line to passenger traffic in July 1959; and it was also used between Over Junction and Newent until it was succeeded by the electric key token system. When Dymock signal box was switched out, the resultant long section between Newent and Ledbury was also worked by the electric key token system.

The line was divided into staff sections, namely Over Junction to Newent; Newent to Dymock; and, after singling of the section, Dymock to Ledbury. Newent and Dymock were both crossing places, where trains could pass (or overtake) one another. Between Over Junction and Newent, Barber's Bridge also had down and up lines (in the form of a loop, with both lines having an adjacent platform), but, so far as is known, it was never operated as a staff station or as a crossing place.

The staff at Newent pose for the photographer *c.* 1904. The signal box mounted on the platform is a Mckenzie and Holland standard pattern. The rodding tunnel is clearly visible.
Colin Maggs

Over Junction

In 1958 the railway bridge spanning the West Channel of the River Severn was rebuilt in an altered position, and the junction between the main and Ledbury lines was adjusted accordingly. As a consequence, the original McKenzie and Holland signal box (of 1884) was replaced by one of unique design on the Western Region: the walls were almost entirely of brick (even the windows had brick surrounds), and, in contrast to the shallow hipped roofs typical of GWR and Western Region practice, the simple pitched roof of the new box was unusually steep. The box had 58 levers, and marked the beginning of the staff section extending to Newent.

Over Junction, seen from the Ledbury Branch, 9th July, 1919. *John Alsop*

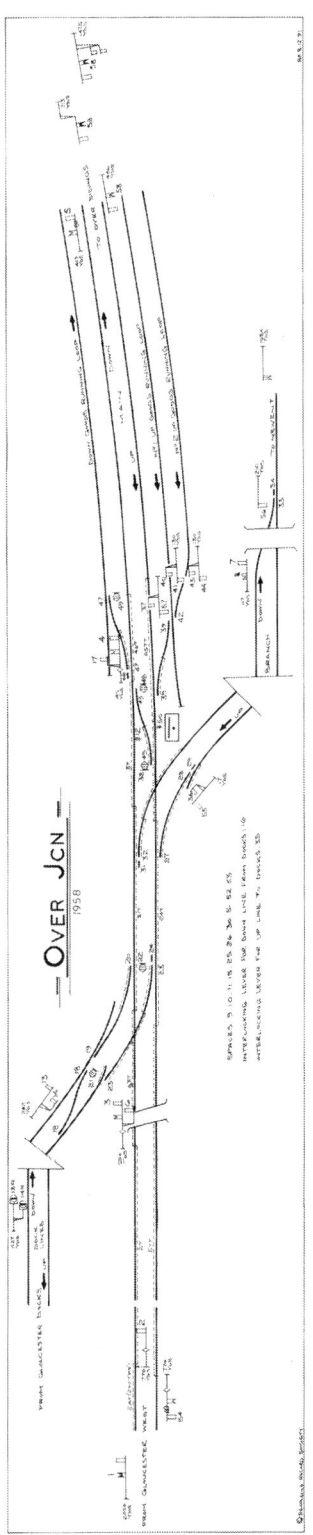

Courtesy the Signalling Record Society

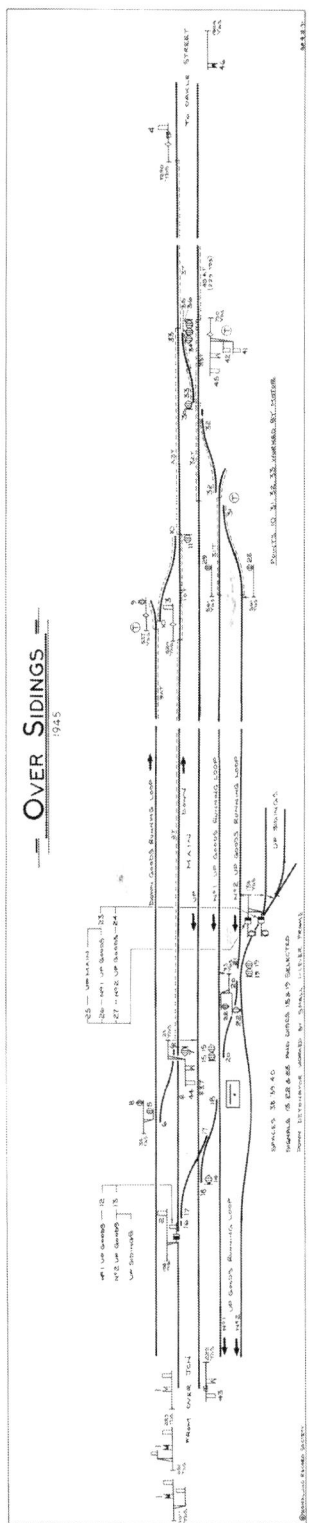

Courtesy the Signalling Record Society

Barber's Bridge

From the opening of the line in July 1885, until July 1898, Barber's Bridge was a block post, with down and up line stop signals. This facilitated the expedient use of the long (8 miles and 34 chains) staff section between Over Junction and Newent: for example, with the use of the staff-and-ticket system, an up train could leave Newent while a preceding up train was still running between Barber's Bridge and Over Junction. Similar arrangements would, *mutatis mutandis*, have applied in the down direction.

In the summer of 1898, when the electric staff system was introduced, the down line at Barber's Bridge was taken out of use (although it remained in situ for several years, and the then redundant down platform survived until after closure of the line); and the station was no longer a block post. This general downgrading of Barber's Bridge may have been in anticipation of the transfer of most through traffic to the Cheltenham and Honeybourne route, although, as we have seen, this new route did not in the event become available for use until August, 1906.

It seems likely that the signal box, which stood towards the north-west end of the down platform, or, more probably, just behind that platform, was of McKenzie and Holland design. After Barber's Bridge ceased, in 1898, to be a block telegraph station, levers in the box may have continued to be used to control the nearer set of points for the goods loop siding. In 1906, this control passed to a ground frame situated a short distance to the west of the up platform; and the more distant access to the loop was controlled by another ground frame located at the Newent end of the loop. These ground frames were released by the key which formed part of the Over Junction to Newent staff or token.

During the time that Barber's Bridge was signalled (from 1885 to 1898), an unusual feature was that after the passage of the last **up** passenger train, which called at the station at 9.22 pm and would have cleared Over Junction at about 9.30, the points were set for the **down** line, the signals were placed at danger, the signal lamps were extinguished, and any subsequent light engines or goods trains travelling in the **up** direction towards Gloucester were authorised to proceed via the **down** line – and to pass unlit signals set at danger – as they continued their journeys towards Over Junction. Normal working was resumed at 7.00 am on the following morning.

Newent

The original signal box at Newent was of the McKenzie and Holland D3 pattern, and was located on the up platform. On Monday, 28th June, 1948, it was taken out of use and succeeded by a GWR box (type 28B), believed to have been removed from Aston Magna (situated between Moreton-in-Marsh and Blockley, on the main line between Oxford and Worcester). This replacement box was installed on the up side of the line and a short distance to the west of the up platform. It was fitted with a GWR vertical tappet frame (with 19 levers, six of which were spares), which would not itself have come from Aston Magna, where there had been a GWR stud frame. After the withdrawal of passenger train services in July, 1959, the box became a ground frame for operating the goods yard points.

The rodding tunnel, through which rods and wires had passed under the platform from the original box, remained clearly visible until the station was finally closed and subsequently demolished. At the far west end of the station area a ground frame (released by the train staff for the Newent to Dymock section) controlled an additional access to the goods yard.

Newent station from the west. The signal box on the up platform was decommissioned in 1948, when a replacement box was provided to the west of the up platform.
Lens of Sutton Association

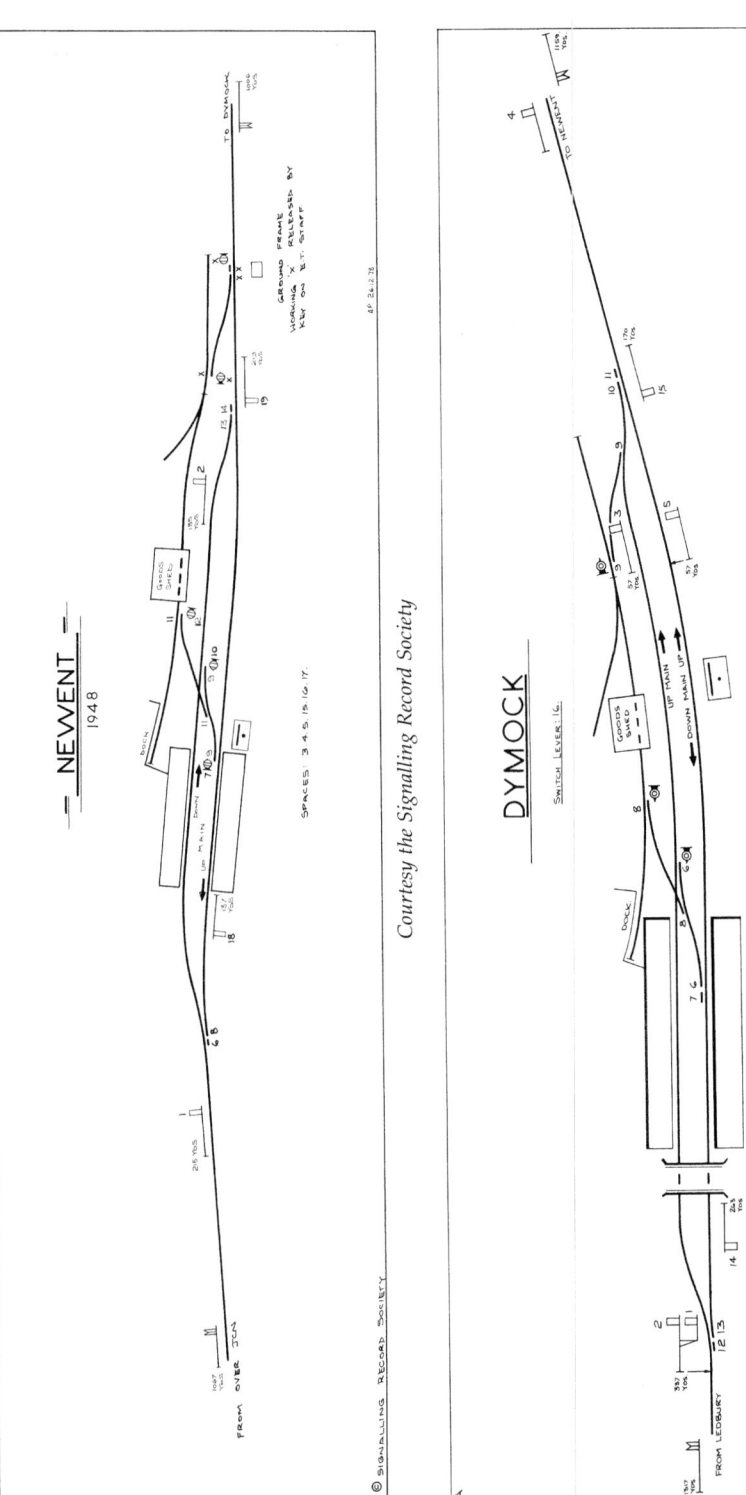

Courtesy the Signalling Record Society

Dymock

Like that originally provided at Newent, the timber signal box at Dymock was of the McKenzie and Holland D3 design. Whereas, however, the Newent box was replaced in 1948, the Dymock box lasted until the end, and its McKenzie and Holland frame remained in use throughout. It had 16 levers. Dymock signal box was situated on the down side of the line and to the south of the station. The box was provided with a switch, which became operative on 6th September, 1930, and from then onwards the line between Newent and Ledbury could be worked as a single 'long section', for which the electric train token system was used instead of the electric train staffs. Also from then onwards, the down line was fully signalled for use by up trains: a bracketed inner home signal was installed to the north of the station, and a starting signal was placed beyond the Gloucester end of the down platform. A facing point lock was added to the down line so as to secure the crossover (leading to the goods yard) when it was in use by up trains.

Dymock station looking south towards Newent. Note the William Clarke style station building. *Tony Harden collection*

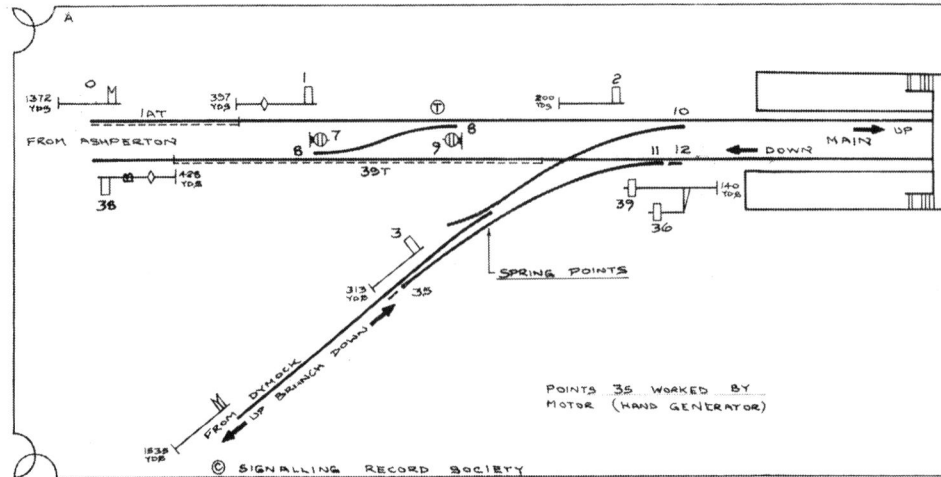

Ledbury Branch Junction

The singling in 1916 of the section between Ledbury and Dymock (above, page 23) occasioned the provision of a small box ('Ledbury Branch') near the junction – situated to the west of Ledbury station and on the up side of the line – which opened on 4th January, 1917. This lasted only until 1925, when the junction with the Worcester and

The double junction at the west end of Ledbury station, with the Gloucester line curving to the left. The double line on the branch became single at Ledbury Branch signal box (not in view here), which was operative between 1917 and 1925. This photograph was taken on 16th June, 1921. *John Alsop*

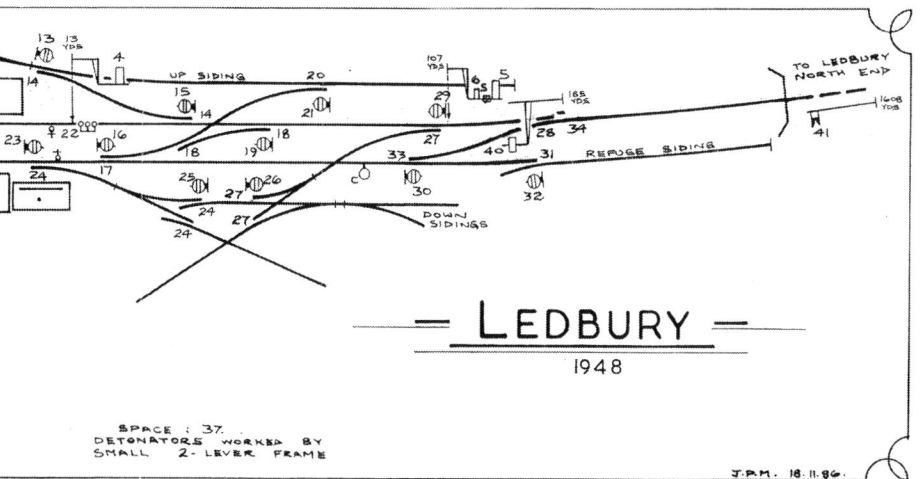

Courtesy the Signalling Record Society

Hereford main line was remodelled and the main signal box at Ledbury assumed responsibility for the whole site. The small Ledbury Branch signal box contained a stud frame provided with six levers, only four of which were in use, presumably to control the points where the single line from Dymock became double preparatory to the junction with the main line, and to control a protective home signal in each direction. Between Ledbury Branch Junction and Dymock the electric train staff system was in operation.

Ledbury

The signal box at Ledbury was situated near the eastern end of the down platform. It was built in 1885 by McKenzie and Holland, and replaced an earlier box dating from the opening of this part of the Worcester and Hereford Railway in 1861. The signalling of the station layout was modified over the years, and for our purposes we may note in particular the changes made (and described above) in connexion with the abolition of Ledbury Branch Junction box. The simplification in 1957 of the junction with the Gloucester line had one interesting consequence in that a new signal, with a route indicator, was installed at the up (eastern) end of the down platform, in order to allow Gloucester line trains, which had necessarily arrived at the down platform, to proceed to the up main line or to the sidings situated to the east of the station.

62 FROM GLOUCESTER TO LEDBURY: THE DAFFODIL LINE

Ledbury station, looking east towards the entrance to Ledbury Tunnel. The tracks behind the up (left hand) platform lead to the locomotive depot. Note the tall signal box, to the left of which is the siding (the 'Gloucester siding') where the Gloucester Branch train could be stabled. *Tony Harden collection*

The box was open continuously. It had 41 levers, and a separate small two lever frame for placing detonators. The box was unusually tall. A curious feature was that the platework of each side of the footbridge between the two platforms was pierced with a large hole, through which the signalman could see the bracket signal (situated at the west end of the down platform) which governed both entry to the main line towards Hereford and access to the branch to Gloucester. (This signal box is still in use, and controls the simplified layout at Ledbury.)

Although Great Western Railway (and later British Railways, Western Region (BR, WR)) pattern signals gradually replaced those of McKenzie and Holland design along the line, the signal box at Dymock retained until the end its McKenzie and Holland frame, which was quite unlike any of the products of the Reading Signal Works of the GWR/WR. The same firm's frame at Newent was removed when the first signal box there was abolished in 1948.

After the complete closure of the section between Dymock and Ledbury on and from 13th July, 1959, the remaining part of the line, between Over Junction and Dymock, was worked as a single section, with an electric train staff which could also be used to release the ground frames situated at Barber's Bridge and at Newent.

Chapter Three

A Journey From Gloucester in Later Years

Trains for Ledbury usually departed from the south side of Gloucester Central station, either from the main westbound platform or, more usually, from the bay on the other side of this platform. The former GWR station had been called simply 'Gloucester', but in 1951 British Railways added the suffix 'Central' to differentiate it from the former LMSR station, which in the same year became 'Gloucester Eastgate'. The Ledbury line trains then ran along the South Wales main line, passing over a viaduct built upon low lying ground between the city and the River Severn, and crossed the East Channel of the River Severn to reach Alney Island, which divided the East and West Channels of the river. Near the western shore of the island a trailing double junction brought to the main line, at Docks Branch Junction, a connexion leading from Gloucester Docks, where there was a complex railway system formerly operated by the Great Western and Midland (later LMSR) Railways.

Next, by means of a bridge renewed in 1958, the South Wales main line crossed the West Channel of the Severn, and almost at once reached Over Junction (1 mile and 39 chains from Gloucester Central), where, again by means of a double, and this time facing, junction, trains to

Gloucester Central station. Ledbury line trains usually left from the bay platform on the right hand side of the picture, but a few used the main down platform in the foreground.
Tony Harden collection

63

64 FROM GLOUCESTER TO LEDBURY: THE DAFFODIL LINE

Over Junction, seen from the east, 9th July, 1919. The photographer is standing on a bridge over the line from Gloucester. To the left the railway goes to Gloucester Docks. Across the bridge over the Severn is the signal box at the start of the Ledbury Branch.
John Alsop

Ledbury diverged to the north-west and passed beneath the main (originally Roman and now the A40 trunk) road leading from Gloucester to Ross-on-Wye. Between the main line and the branch was situated Over Junction signal box. The Ledbury line became single and rose slightly in order to take up the course of the Hereford and Gloucester canal.

Just before the railway joined the line of the canal, there was on the right hand side a large basin for use by waterborne traffic, and, a little further to the east, a deep lock through which canal boats could gain admission to the West Channel of the river, and vice versa. During the years 1998 to 2000 this basin was successfully restored and extended by the Herefordshire and Gloucestershire Canal Trust, as one of its many initiatives being taken with a view to the eventual reopening of the canal for its entire length between Hereford and Gloucester.

The railway curved to the east of Lassington Woods and then followed the course of the canal, which had itself here run closely to the west of the River Leadon. There having been no locks along this section, the railway was mainly level before a slight inclination was met close to the small settlement of Rudford, where there had been a single lock. To the west of Rudford lay the village of Tibberton, and also Tibberton Court, the home of the Price family, who had been so influential in the transport planning of the district.

The line was crossed by a skew road bridge carrying the road (B4215) leading from Highnam (west of Over) to Newent, Dymock and Ledbury, and then at once entered the station of Barber's Bridge, 5 miles and 37 chains from Gloucester Central.

Barber's Bridge

The station here was named after a nearby bridge spanning a small tributary of the River Leadon. The apostrophe was omitted from the large nameboard situated on the platform, and also from tickets to and from the station, but was included in timetables and maps. (The apostrophised form is the more correct.)

At Barber's Bridge were two platforms, and upon the up (southbound) platform stood a station building of typical William Clarke design, as described above (pages 27-29). A corrugated-iron shed located on this platform was used to accommodate parcels and other small items.

In 1898 the the running loop and the down platform had been taken out of use. Although it became much overgrown, the marooned platform remained in situ until after the closure of the line. Two ground frames, one near the west end of the up platform and the other further to the west, controlled access to the goods yard loop siding.

Barber's Bridge station in 1919, seen from the overbridge at the south end of the station. The former down line has been removed, and the down platform is redundant. On, or just behind, it had stood a signal box, by now abolished. The points on the running line are controlled by two hutted ground frames, one of which is at the far end of the main (Gloucester bound) platform. Note the absence of any running line signals.

John Alsop

To the north of the passenger station, and on the east side of the line, was a small goods yard, complete with a crane capable of lifting 3 tons. The yard included the depot of the local family firm of Charles Teague. Charles P. Teague was a highly respected local figure, a coal merchant, carpenter, wheelwright and builder. He owned one, or perhaps more than one, private owner wagons, built by the Gloucester Railway Carriage and Wagon Company, and painted a deep russet colour with white lettering.

In the later years of the line, Barber's Bridge was staffed only during the middle of the day, and at other times was treated as a halt.

On to Malswick

After Barber's Bridge the canal, and consequently the railway, veered away from the river, and took up a north-westerly course through open country to reach Newent. About a mile and a quarter before Newent the railway crossed by an overbridge a lane ('Hook's Lane') running from the B4215 road towards Upleadon, and then served Malswick Halt, 8 miles and 47 chains from Gloucester Central. Opened on 1st February, 1938, and situated on the west side of the line, the halt was of timber construction and stood upon an embankment. Access was by means of a series of wooden steps leading upwards from the lane which the railway had just crossed. Upon the platform was a wooden shelter, at the front of which was a small valanced canopy. The halt was lit by an oil lamp mounted at the top of a tall post, although in later years evening trains from Ledbury usually did not call at Malswick. Responsibility for Malswick Halt belonged to the station master at Newent.

Newent

The railway now continued to the important station of Newent, situated at 9 miles and 73 chains from Gloucester Central. This was a crossing station, provided with down and up platforms 300 feet in length. The main station building, again of William Clarke design, stood on the down (towards Ledbury) platform. On the up (towards Gloucester) platform stood a small timber built shelter for the use of waiting passengers. Curiously, one of the outdoor benches placed on this platform was of the Midland Railway rustic cast-iron pattern.

A JOURNEY FROM GLOUCESTER IN LATER YEARS

In this typical 1950s scene, prairie tank locomotive No. 4573 enters Barber's Bridge station with a train from Gloucester. A telegraph pole stands on the site of the former down loop, and beneath the flowers on the right hand side slumbers the down platform, which survived up to and after closure of the line. Note the large brackets supporting the canopy of the William Clarke building. *Tony Harden collection*

Malswick Halt, looking towards Newent. The platform shelter is similar to those of the other halts along the line. *Michael Clemens*

Newent station, with a Gloucester bound train entering the up platform, upon which the signal box is just visible. *John Alsop*

The goods yard was located at the west end of the station and was approached by a typical GWR layout arrangement: a trailing connexion from the up loop line led to the yard, and a single slip from this connexion provided a crossover between the two running lines (see the diagram). The yard itself contained several sidings; a goods shed capable of holding three four wheeled vehicles for loading and, or, unloading; and a provender store. John Meates, a local coal merchant, had an office situated in the yard. A crane capable of lifting 5 tons and a weighbridge were provided. The yard was also home to a private owner wagon belonging to W. R. Field. After the closure of the line in 1959 to passenger traffic, goods trains continued to serve Newent, but in 1960 the goods shed was demolished, and from then on general goods which needed to be covered were accommodated in the provender store.

To the west of the goods yard, a headshunt provided access for coal wagons to serve a pumphouse which delivered water for piping to Gloucester. This water was taken from the nearby Ell Brook, a tributary of the River Leadon.

The first McKenzie and Holland signal box had been located close to the west end of the up platform, and, until closure and beyond, the rodding tunnel (through which the point rods and signalling wires passed) remained clearly visible as an arched gap in the wall of the platform. In 1948 this box was replaced by another (possibly transferred from Aston Magna), and was newly sited a short distance to the west of the up platform. At the far west end of the station area, a small ground frame, released by the train staff, controlled an additional access to the goods yard.

A JOURNEY FROM GLOUCESTER IN LATER YEARS 69

A wet day at Newent, tooking west towards Dymock. The McKenzie and Holland signal box stands on the up platform. The station building on the down side is again by William Clarke. Beyond the down platform the extensive goods yard is visible.
Lens of Sutton Association

A train from Ledbury enters the up platform at Newent. *Tony Harden collection*

Four Oaks Halt, situated on a level section in a cutting at the summit of the line, between Newent and Dymock. The view here is northwards, towards Dymock. Note the very high position of the lamp. *M. Hale/Great Western Trust*

The Gable, and Four Oaks Halt

Shortly after leaving Newent, the line began to ascend at a gradient of 1 in 80 in order to cross the ridge beneath which the canal passed in the Oxenhall Tunnel. The climb was succeeded by a brief level section in a deep cutting in which Four Oaks Halt (12 miles and 6 chains from Gloucester Central) was situated on the down side of the line. Its timber edged platform had a cinder surface and it was provided with a small wooden shelter, on the north side of which a tall post supported an oil lamp. Responsibility for the halt lay with the station master at Dymock.

The railway now descended from the summit at the same inclination by which it had risen, that is, at 1 in 80. Running through picturesque surroundings, with on the west side parts of the 'Golden Triangle' being visible, it rejoined the line of the canal after the latter had emerged from the Oxenhall Tunnel and had passed Boyce Court. The line now approached the west side of the village of Dymock, on the east side of which flowed the River Leadon.

Dymock

Dymock station (at 13 miles and 66 chains from Gloucester Central) was a passing place, and indeed had been the southern end of the double

A JOURNEY FROM GLOUCESTER IN LATER YEARS 71

Dymock, looking north towards Ledbury, with the staff on parade. The station is neat, but looks bare in comparison with its later sylvan appearance *John Alsop*.

track section which had run southwards from Ledbury before it was singled during the First World War. Dymock station was treated as the northern limit of the Newent Railway, and, just as at Barber's Bridge and at Newent stations, the initials 'N.R.' could be seen in the spandrels of the canopy of the main station building. Once again this building (this time on the up side) was of the pleasing William Clarke design; and once again there was on the other (this time down) platform a small timber built shelter for waiting passengers.

Situated south of the station, and on the down side, the McKenzie and Holland signal box had 16 levers. The goods yard was also situated to the south of the passenger station, and was almost a mirror image of that at Newent: whereas the yard at Newent was on the down side, that at Dymock was on the up side. The layouts were similar in that access from the running lines was by means of a trailing connection, in Dymock's case from the down line. This connection crossed the up line by a diamond crossing, which also had a single slip, as at Newent, to provide a crossover between the two running lines. Also as at Newent, a 5 ton crane was installed. Merchants having private owner wagons stationed at Dymock included J. Barnett and Co., and A. E. Griffiths, the latter dealing in coal, bricks, lime and salt.

Dymock seen from the road bridge at the north end of the station. The track layout here is almost a mirror image of that at Newent. To enable up (towards Gloucester, on left) passenger trains to use the down line, a facing facing point lock at the crossover and a starting signal, beyond the signal box, were installed. *Lens of Sutton Association*

Although the main building was situated on the up side of Dymock station, it was the down platform line which was treated as bi-directional and which was signalled accordingly. Presumably this was because the down line was straight, whereas the up line took the form of the loop. This arrangement meant that when the signal box was switched out passengers wishing to join a train travelling in the direction of Gloucester had to use the board crossing to gain the down platform, just as they did when they wished to board a train for Ledbury. In fact, none of the intermediate stations on the line had a footbridge to connect the platforms.

Greenway Halt and the approach to Ledbury: the Town Halt

After leaving Dymock, the railway again closely followed the line of the canal, and was soon reunited with the valley of the River Leadon. The line ran through a verdant landscape enriched by orchards and small streams. At 15 miles and 25 chains from Gloucester Central was situated (on the up side), Greenway, which served a tiny settlement of that name, and, a little more distantly, the hamlet of Broom's Green. This halt was built on the site of the original up track of the double line between

A JOURNEY FROM GLOUCESTER IN LATER YEARS

Greenway Halt was built on the site of the former up line between Ledbury and Dymock. The view here is southwards, towards Dymock. The bridge span over the original double track formation is clearly seen. Again, the lamp is in a very high position.

M. Hale/Great Western Trust

A small spectator watches the careful transfer of a perambulator at the diminutive Greenway Halt.
E. Wilmshurst/Rail Photoprint

Dymock and Ledbury, and so was a clear indication that there was no intention to reinstate two line working between those two stations. Greenway Halt consisted of a simple cinder surfaced and timber edged platform, upon which stood a wooden shelter having a narrow canopy at the front. To the south of this shelter an oil lamp was mounted upon a tall post. In common with Four Oaks Halt, Greenway Halt was under the control of the station master at Dymock.

Shortly after passing Greenway, the railway crossed the River Leadon, which at this point constituted the boundary between Gloucestershire and Herefordshire; and the train now arrived at the site of the basin which had for many years been the northern terminus of the canal. From this basin no fewer than five locks had been necessary to raise the level of the canal to that of the wharf in Ledbury itself at Bye Street (near the centre of the town), which had been finally reached in 1841. The railway followed the same course and correspondingly rose at a gradient of 1 in 72, and in the process had obliterated all five locks. The railway passed below Bye Street, and immediately came to Ledbury Town Halt (18 miles and 20 chains from Gloucester Central), which was situated on the up side and again on the site of the former up line. Similar to Greenway, the halt again had a simple cinder platform with a small shelter, in this case built of corrugated-iron. Lighting was originally by means of oil lamps, but electric light was later installed. A

Ledbury Town Halt oil lamps flank the corrugated-iron shelter. Underneath the bridge, is the small shed where paraffin was stored. *Lens of Sutton Association*

small hut positioned under the Bye Street bridge contained the paraffin supply for use when the halt was lit by oil. The station master at Ledbury was responsible for the Town Halt.

For some time after the opening of this halt, tickets could be purchased at the bakery and grocery shop of Mr Hodges, at 42 Bye Street, who was an official agent of the GWR. Thereafter passengers from the halt paid at Ledbury or at Dymock for their journeys, or bought their tickets from the guard.

Ledbury

The line then rose steeply, at a gradient of 1 in 64, to reach the junction, just to the west of Ledbury station, with the Worcester and Hereford main line. From 1916 until 1957, the branch became double in order to effect a double junction with the main line, but in the latter year the layout was simplified: the double junction was removed, and trains from the branch were then able to reach only the down main line and the down platform.

Ledbury, 18 miles and 79 chains from Gloucester Central, was an important station. There were (and are) two platforms, serving the down and up main lines. The main building, signal box and goods yard were all situated on the down (south) side, while the up (north) side

General view of Ledbury station, looking west towards Hereford. The large circular holes in the platework of the footbridge enabled the signalman to see the bracket signal at the west end of the station. *Tony Harden collection*

platform held a subsidiary but still substantial brick and canopied waiting room. Behind this platform were situated the small locomotive depot and a siding. Towards the east end of the layout the double line became single preparatory to entering the steeply graded Ledbury Tunnel (1,323 yards), at the far end of which was Ledbury North End signal box and loop. From 1957 the incoming train from Gloucester could stand at the down platform and await time for the return journey, but would need to vacate the platform if the line was needed for the passage of a main line down train: the branch train could then be stabled in a layover siding (the 'Gloucester siding') located at the east end of the station.

This compact, but well appointed, station was the terminus, for most of its history, of the Gloucester to Ledbury railway.

Here, seen at Dymock, is the last ever down passenger train over the branch. On 11th July, 1959, engine No. 3203 heads the 6.24 pm train from Gloucester to Ledbury.

Hugh Ballantyne/Rail photoprints

Epilogue

At Newent passengers wait to join a Gloucester bound train. Newent signal box is just visible beyond the shelter, July 1912.
John Alsop

The story of the railway which ran between Over Junction and Ledbury is one of metamorphosis. The Hereford and Gloucester Canal made way for the railway, and today the course of the railway is being reclaimed for the restoration of the canal.

The local companies which planned the two railways which would take over the route of the canal and which would connect Gloucester and Ledbury envisaged the development of traffic which would enhance the economy and the amenity of the district: there might also be scope for the use of the railway as part of a through route in the larger transport network. The Great Western Railway, which adopted and implemented much of the scheme, inverted these priorities: for the GWR the chief value of the line would lie in its provision of a new through route from Gloucester to Worcester and to the West Midlands, and the carriage of local passengers and goods would be, whilst still important, a secondary function. The section between Dymock and Ross-on-Wye for which Parliamentary powers had been obtained was not necessary for the GWR vision, and it was never built.

After the opening of the line in 1885, the dual role contemplated by the GWR was fulfilled for just over two decades. When, however, a more direct and faster route from Gloucester via Cheltenham and Honeybourne to Worcester and to Birmingham became fully available in

1908, through freight traffic was transferred away from the Ledbury line, which in consequence was relegated to the status of a mainly local line, largely confined to the service of the district through which it ran.

In this service, the line was quite effective, and had some success in resisting the gradual modal shift from rail to road which characterised the period between the two world wars. For passengers the only inconveniently situated station was that at Ledbury, and this disadvantage was overcome by the provision of Ledbury Town Halt in 1928; and in 1937 and 1938 three other halts – Greenway, Four Oaks, and Malswick – were created to serve small communities. To its credit, the GWR also cultivated excursion traffic during this period; and the introduction in 1940 of diesel railcars to replace steam trains led to valuable operating economies. Petrol shortages during the Second World War retarded the growth of private car usage, as did petrol rationing, which continued until 1950.

Nevertheless, the demise of the line was probably inevitable: against it were ranged the rise of commercial and private road transport, and – in a less visible but more lethal way – the management of British Railways Western Region, which during the 1950s evinced an implacable zeal for the elimination of secondary and branch railways and of wayside stations. There was no attempt to bring the timetable up to date to accommodate changing economic and social patterns of travel: on the contrary, the level of passenger services was ruthlessly reduced in 1958, in preparation for their complete termination in 1959. More than 60 years later, one can only regret the myopic approach which led to the end of this charming byway in deeply rural England. Our consolation must be in the present determined attempts being made by the Herefordshire and Gloucestershire Canal Trust to re-instate the canal with which our story began, and to replace the abandoned railway with a ribbon of bright water.

Bibliography

Ashworth, B., *The Last Days of Steam in Gloucestershire*, revised edition, Amberley Publishing, 2009.
Bartlett, S., *Gloucester Locomotive Sheds*, Pen and Sword Transport, 2019.
Beale, G., 'The 'Standard' Buildings of William Clarke', in *British Railway Journal* No. 8, Wild Swan Publications, 1985, pages 266 – 276.
Bick, D. E., *The Hereford and Gloucester Canal*, Poundhouse, Newent, 1979.
- *The Hereford and Gloucester Canal*, second edition, Oakwood Press, 1994
- *The Hereford and Gloucester Canal*, new edition, Oakwood Press, 2003. (N.B. All three editions include a contribution by Norris, J. E., about the Gloucester to Ledbury Railway.)

Christiansen, R., *A Regional History of the Railways of Great Britain*, Volume 13, Thames and Severn, David and Charles, 1981.
Clarke, R. H., *An Historical Survey of Selected Great Western Stations*, Volume Two, Oxford Publishing Co., 1979. [For Dymock station.]
- *An Historical Survey of Selected Great Western Stations*, Volume Three, Oxford Publishing Co., 1981. [For Ledbury station.]

Dale, P., *Gloucestershire's Lost Railways*, Stenlake Publishing, 2002.
Goode, C. T., *The North Warwickshire Railway*, Oakwood Press, 1978.
Jenkins, S. C., and Parkhouse, N., 'The Gloucester to Ledbury Branch', in *Railway Archive* No. 36, Lightmoor Press, September, 2012, pages 3 – 38.
Judge, C. W., *The History of the Great Western A. E. C. Diesel Railcars*, Oxford Publishing Co., 1986.
Maggs, C. G., *The Branch Lines of Gloucestershire*, Amberley Publishing, 2007.
Mitchell, V. and Smith, K., *Worcester to Hereford, including the Branches to Leominster and Gloucester*, Middleton Press, 2004, reprinted 2012.
Norris, J. E., 'The Gloucester and Ledbury Branch', in *The Railway Magazine*, Volume 104, 1958, pages 228 to 232.
- Contributions to each edition of Bick, D. E., *The Hereford and Gloucester Canal* (above).

Oppitz, L., *Lost Railways of Herefordshire and Worcestershire*, Countryside Books, 2002.
Parkhouse, N., *British Railway History in Colour*, Volume 1, West Gloucester and Wye Valley Lines, second edition, Lightmoor Press, 2018.
Parr, H. W., *The Great Western in Dean*, second edition, David and Charles, 1971.
Pope, I., *Private Owner Wagons of Gloucestershire*, Lightmoor Press, 2006.
Postle, D., *From Ledbury to Gloucester by rail*, Amber Graphics, Ledbury, 1985.
- 'The Ledbury to Gloucester Branch', in *Steam Days*, No. 158, October, 2002, pages 594 – 606.

Potts, C. R., *An Historical Survey of Selected Great Western Stations*, Volume Four, Oxford Publishing Co., 1985. [For Newent Station.]
White, H. P., *Forgotten Railways*, David St John Thomas, 1986.
Wood, G., *Railways of Hereford*, Gordon Wood in conjunction with Kidderminster Railway Museum, 2003.

Index

Alney, Island of, 7, 14, 63
Appleby and Lawton, 19
Ashperton Tunnel, 12
Aston Magna, 57, 68
Auto-train, 9, 30, 41
Aylestone Tunnel, 12

Banking engine, 44, 52
Barber's Bridge, 19, 20, 21, 26, 28, 46, 56, 65, 67; Simplification, 21, 52
Beeching, Dr Richard, 24
Birmingham, 23, 31, 35, 38, 41, 42, 77
Block Section, Barber's Bridge, 56; Ledbury to Newent, 38, 59; Over Junction to Dymock, 47; Over Junction to Newent, 21-22
Board of Trade, 20
Brindley, James, 8
British Railways, 44, 62, 78

Cheltenham, 33, 38, 41
Cheltenham and Honeybourne line, 22, 43, 56, 77
Clarke, William, 19, 20, 27, 59, 65, 66, 67, 71
Closure, end of goods services, 51; end of passenger services, 42, 50, 68; Ledbury to Dymock, 24
Clowes, Josiah, 8, 11, 26
Coal, 8, 11, 12, 44, 66, 68, 71
Colwall Tunnel, 42, 44
Cottman, Mr G., 19

Daffodil Line, 32, 45
Daffodils, 32, 45
Double track, 19, 53; Conversion to single track, 23, 53; Greenway Halt built on, 72, 73; Provision for doubling Newent Railway, 19
Dymock, 8, 15, 16, 20, 23, 26, 29, 43, 47, 70, 77; Part-time signal box, 38; Signalling, 52, 58, 59

First World War, 23, 32, 71
Forest of Dean, 5, 14; Coalfield, 8
Four Oaks Halt, 24, 26, 29, 33, 70, 78
Gloucester, 5, 7, 12, 21, 23, 31, 38, 42; Central Station, 20, 49, 63; Docks, 7, 43, 63
Gloucester Horton Road, 50, 51
Gloucester to Birmingham through service, 31, 38, 41

Gooch, Sir Daniel, 16
Goods traffic, 42-47
Gradients, 26
Grange Court Junction, 12, 16
Great Western Railway, 14, 16, 29, 32, 46, 62, 77; Acquisition of line, 21; Amalgamation of WMR, 13; Country Lorries, 46, 47; Newent alternative routes from Paddington, 32
Greenway Halt, 24, 29, 33, 72-74, 78

Hall, Richard, 7, 8
Henshall, Hugh, 8
Hereford, 5, 7, 12
Hereford and Gloucester Canal, 6, 7-8, 11-14, 64, 77; Closure between Ledbury and Severn, 19; Conversion to railway, 19, 20; GWR takes control, 13; Hope for railway purchase, 12; Newent Route, 11; Operating agreement with WMR, 13; River Leadon route, 8
Hereford and Gloucester Canal Navigation Company, 8, 14
Hereford and Gloucester Canal Trust, 64, 78

Kidderminster, 38, 47

Ledbury, 8, 12, 26, 74, 77
Ledbury Basin, 11, 26, 52, 74
Ledbury Branch Signal Box, 23, 60
Ledbury Junction, 1, 26, 52, 60; Double junction, 23, 75
Ledbury Station, 1, 2, 9, 10, 17, 22, 31, 47, 62, 75-76; Signal Box, 17, 61, 62, 76
Ledbury Town Halt, 10, 23, 25, 26, 29, 33, 47, 74, 78
Ledbury Tunnel, 42, 44, 52, 76
Leominster, 7
London Midland and Scottish Railway, 44, 63
Loss of traffic, 22, 44

Malswick Halt, 24, 29, 33, 35, 66, 67, 78; Operation as request stop, 35
Malvern Hills, 5
McKenzie and Holland, 52, 54, 56, 57, 59, 61, 68, 71
Midland Railway, 32, 42, 44, 66

Nationalisation, 14, 24
Newent, 8, 20, 23, 25, 26, 28, 29, 31, 35, 47, 49, 53, 66, 69; Alternative routes from Paddington, 32; Aston Magna signal box, 57, 68; Coal mining, 11, 44; Goods shed demolished, 68; Signalling, 52, 57, 58
Newent Railway, 15, 19, 29, 71

Over, 8
Over Junction, 15, 20, 26, 63, 77; Signalling, 52, 54-55
Oxenhall Tunnel, 11, 26, 47, 70; Conversion to railway, 16
Passenger Services, first, 20, 21; timetables, 31-42
Price, William, 13
Proposed Railways, 12-16

Railcar, 2, 10, 24, 25, 28, 29, 35, 38, 48, 50, 78; serviced at Cheltenham, 50
Rich, Colonel Frederick Henry, 20
River Leadon, 7, 8, 26, 64, 65, 70, 72, 74
River Severn, 7, 14, 43, 54, 63
Road traffic, competition, 24; petrol rationing, 24, 78
Ross and Ledbury Railway, 15, 16, 19
Ross Ledbury and Gloucester Railway, 14-15
Ross-on-Wye, 5, 12, 14, 16, 63, 77

Second World War, 24, 35, 78
Severn Commissioners, 14
Single Line, 19, 23, 52; working, 53, 56, 59, 62
South Wales, 7, 12, 46
Sunday Services, 32, 45

Use as diversionary route, 44

Weight and speed restrictions, 51-52
West Midlands, 5, 45, 77
West Midlands Railway, 13
Whitworth, Robert, 11
Worcester, 16, 23, 31, 42, 43, 77
Worcester and Hereford Railway, 12, 26, 60, 61
Worcester, Dean Forest and Monmouth Railway, 14
Worcester, End of through services from Gloucester, 32
Wye Valley Railway, 12